CAMBRIDGE LIBRARY COLLECTION

Books of enduring scholarly value

Cambridge

The city of Cambridge received its royal charter in 1201, having already been home to Britons, Romans and Anglo-Saxons for many centuries. Cambridge University was founded soon afterwards and celebrates its octocentenary in 2009. This series explores the history and influence of Cambridge as a centre of science, learning, and discovery, its contributions to national and global politics and culture, and its inevitable controversies and scandals.

A Handbook to the Geology of Cambridgeshire

Beneath Cambridgeshire's towns, villages, farmland, hills, fens and waterways lie the rocks that display a variety of geological landscapes. Basement rocks are buried under sandy deposits from ancient tropical seas. The rising and tilting of the land due to large-scale movements permitted water flows that produced gradual alterations. Glaciation, erosion and dramatic variations in climate all wrought more rapid changes. The consequences of these processes are revealed in this scholarly 1897 account of the geology of Cambridgeshire, which integrated the latest research then available. Proceeding from the most ancient to the most recent beds, systematic consideration is given to the features, distribution and modes of formation, as well as the economic implications of the various strata. Discussions of palaeontology, including detailed lists of site-specific fossils, and of water supply are also provided. An appendix lists maps, memoirs and other publications of H.M. Geological Survey from 1814 to 1897.

Cambridge University Press has long been a pioneer in the reissuing of out-of-print titles from its own backlist, producing digital reprints of books that are still sought after by scholars and students but could not be reprinted economically using traditional technology. The Cambridge Library Collection extends this activity to a wider range of books which are still of importance to researchers and professionals, either for the source material they contain, or as landmarks in the history of their academic discipline.

Drawing from the world-renowned collections in the Cambridge University Library, and guided by the advice of experts in each subject area, Cambridge University Press is using state-of-the-art scanning machines in its own Printing House to capture the content of each book selected for inclusion. The files are processed to give a consistently clear, crisp image, and the books finished to the high quality standard for which the Press is recognised around the world. The latest print-on-demand technology ensures that the books will remain available indefinitely, and that orders for single or multiple copies can quickly be supplied.

The Cambridge Library Collection will bring back to life books of enduring scholarly value across a wide range of disciplines in the humanities and social sciences and in science and technology.

A Handbook to the Geology of Cambridgeshire

For the Use of Students

FREDERICK RICHARD COWPER REED

CAMBRIDGE
UNIVERSITY PRESS

CAMBRIDGE UNIVERSITY PRESS

Cambridge New York Melbourne Madrid Cape Town Singapore São Paolo Delhi

Published in the United States of America by Cambridge University Press, New York

www.cambridge.org
Information on this title: www.cambridge.org/9781108002394

© in this compilation Cambridge University Press 2009

This edition first published 1897
This digitally printed version 2009

ISBN 978-1-108-00239-4

CAMBRIDGE GEOLOGICAL SERIES.

A HANDBOOK

TO THE

GEOLOGY OF CAMBRIDGESHIRE.

London: C. J. CLAY AND SONS,
CAMBRIDGE UNIVERSITY PRESS WAREHOUSE,
AVE MARIA LANE,

AND

H. K. LEWIS,
136, GOWER STREET, W.C.

Glasgow: 50, WELLINGTON STREET.
Leipzig: F. A. BROCKHAUS.
New York: THE MACMILLAN COMPANY.
Bombay and Calcutta: MACMILLAN AND CO., LTD.

N.W.   S.E.

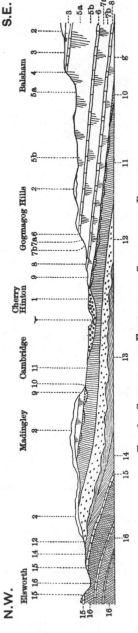

FIG. 1. SECTION FROM ELSWORTH BY CAMBRIDGE TO BALSHAM.
(After Prof. T. McK. Hughes.)

Vertical scale about 800 ft. to an inch.   Length of section about 21 miles.

ϒ   Recent alluvium.
1. River gravels.
2. Glacial Drift.
3. Zone of *Micraster cor-testudinarium* (Upper Chalk).
4. Chalk Rock or Zone of *Heteroceras reussianum.*
5 a. Zone of *Holaster planus.*
5 b. Zone of *Terebratulina gracilis* and Zone of *Rhynchonella Cuvieri.*
6. Melbourn Rock at base of Zone of *Rh. Cuvieri.*
7 a. Zone of *Belemnitella plena.*
7 b. Zone of *Holaster subglobosus* (Cherry Hinton Chalk).
8. Burwell Rock at base of Zone of *Hol. subglobosus.*
9. Chalk Marl or Zone of *Ammonites varians.*
10. Cambridge Greensand at base of Zone of *Am. varians.*
11. Gault.
12. Lower Greensand.
13. Kimeridge Clay with thin bands of limestone and septarian nodules.
14. Ampthill Clay including the Upware limestones (Corallian).
15. Elsworth and St Ives Rocks (Lower Calcareous Grit).
16. Oxford Clay with thin bands of limestone.

[*Frontispiece*

A HANDBOOK

TO THE

GEOLOGY OF CAMBRIDGESHIRE,

FOR THE USE OF STUDENTS

BY

F. R. COWPER REED, M.A., F.G.S.

ASSISTANT TO THE WOODWARDIAN PROFESSOR OF GEOLOGY.

CAMBRIDGE:
AT THE UNIVERSITY PRESS.

1897

Cambridge:

PRINTED BY J. AND C. F. CLAY,
AT THE UNIVERSITY PRESS.

PREFACE.

THE earliest paper on Cambridgeshire Geology was published more than a hundred years ago, but the first important general description was given in 1861 in the lecture by Professor Sedgwick on "The Strata near Cambridge and the Fens of the Bedford Level." It was first reported in the local press but afterwards reprinted with a supplement and diagram sections. Two years previously Mr Lucas Barrett, assistant to Prof. Sedgwick, had published with the aid of the latter a map of the neighbourhood of Cambridge. Since then a multitude of geologists have studied the stratigraphy and palæontology of the district and have added immensely to our knowledge by their detailed investigations, but they have only tended to establish the general accuracy of the broad outlines sketched in by Professor Sedgwick.

Amongst the earliest and most energetic workers at the local geology was Prof. H. G. Seeley, whose careful and energetic research in this district is represented not only by numerous papers in various scientific journals but by the arrangement of a considerable part of the local collections in the Woodwardian Museum.

In 1872 Mr A. J. Jukes-Browne, who for many years has been issuing the valuable results of his geological

work in East Anglia, published a revised version of Mr
Barrett's map; and in 1875 Prof. T. G. Bonney, who had
been requested to re-edit Prof. Sedgwick's lecture of 1861,
brought out his excellent manual of "Cambridgeshire
Geology," which since then has been the standard book
on the subject for students.

The Memoir of the Geological Survey on the southern
part of the county issued in 1881 was followed in 1893
by that on the northern portion; and these with the
original articles and papers of which a list is given in the
appendix must be consulted for details.

Owing to the recent publication of Sheets 9 and 12
of the inexpensive Index Map of the Geological Survey
it has been considered unnecessary to increase the cost of
the present handbook by the addition of a geological
map, since these Sheets include the whole of Cambridgeshire and portions of the adjoining counties.

The increasing number of geological students in the
University and their constant request for a concise
account of the local geology embodying the results of
the most recent research have led me to prepare this
little book. I am especially indebted to Prof. T.
McKenny Hughes for invaluable assistance and advice,
and to Prof. Bonney for his kind permission to make
free use of his manual on the local geology. My thanks
are also due to Mr H. Woods for much information
about the Cretaceous beds.

CONTENTS.

LIST OF FIGURES.

INTRODUCTION.

THE Cambridgeshire area is full of interest to the geologist not only on account of the variety of the strata which range from the Oxford Clay to recent fluviatile deposits, but also because of the peculiar local development and relations of some of the formations. We may mention for instance the argillaceous representative of the Corallian limestones of other districts, the Upware coral reef, the Cambridge Greensand, and the river-terraces.

Physical features. There are four natural regions depending on geological structure into which the County of Cambridgeshire may be divided. These are (1) the chalk hills and plateau in the south and east; (2) the high ground in the west; (3) the median valley of the Rhee, and the Cam, and (4) the Fenland in the north. There is no definite line of demarcation between the last two regions, for the river-valley widens out gradually and passes imperceptibly into the Fenland.

The first region comprises the whole range of chalk hills from Royston in the south-west to Newmarket in the north-east, and from it flow down the main tributaries of the Cam and Ouse, cutting their way through the face of the escarpment. In many parts the typical scenery of a chalk country is displayed, as for instance round Royston, where the well-known rolling outlines of chalk downs

R.   1

covered with short smooth turf present a marked contrast
to the level plain in the northern part of the county.
Over a large area, however, post-tertiary deposits cover
the solid rock and the characteristic features are obscured.
The highest ground occurs in the eastern region. Thus
at Tharfield the escarpment is 500 feet high, whereas
its average height is about 300 feet in the Gog Magog
Hills.

The second region is an irregular hilly district composed partly of outliers of Lower Chalk with long
promontories projecting eastwards, the whole capped by
an extensive layer of Boulder Clay. In the south-west
the Lower Greensand rises into picturesque hills of some
importance, as near Sandy. The Jurassic clays lie still
further to the west and form a plain except where patches
of drift or gravel cause gentle elevations or ridges. This
is particularly the case in the northern part where the
plain merges into the Fenland.

The third region comprises the valley and alluvial plain
of the Cam, the Rhee, and their tributary streams. It is
composed chiefly of Gault and Chalk Marl with a covering
of gravel along the present and ancient river-courses.
The town of Cambridge is situated on the Gault in the
middle of this valley. The portion of this region composed
of Chalk Marl is gently undulating but the Gault areas are
mostly very flat. The lines of gravel give rise to ridges
and slightly elevated ground, and the three river-terraces
are well developed.

The fourth region consists of peat-land in the south
and silt-land in the north. The flat open treeless character
of the landscape where peat prevails is well known. A
long tongue of the Fenland projects southward to include
Burwell Fen. To the west of the Cam the Ouse forms

the southern boundary of this region between St Ives
and Thetford. A few scattered 'islands' rise above the
general dead level, the most important of which are the
Isle of Ely formed of Kimeridge Clay capped by Lower
Greensand and Boulder Clay; the raised ground round
March which consists of Boulder Clay and marine gravel;
and the inlier of Jurassic rocks and Lower Greensand
composing the Upware ridge.

A great expanse of level ground stretches from
Cambridge in the south to the Wash in the north—a
distance of about forty-five miles, and the whole Fenland
forms the largest plain in England, having an area of about
1300 square miles. The Wash was once coextensive with
the Fenland and the whole silting up of this vast area has
taken place within recent geological times[1]. In fact the
Fenland is an enormous complex delta formed by several
rivers, flowing chiefly from the south-west into a large
shallow bay. At the landward end they deposited their
loads of detrital material which stretched out as muddy
flats for miles into the sea. The tides brought in fine
silt which was thrown down against the outward face of
the delta and gradually formed banks which rose above
high-water mark. Over these low-lying lands forests and
peat-bogs ultimately spread. The process by which the
land encroaches in this way on the sea and by which the
sea itself is silted up is still going on in the Wash though
at a slower rate than in past times, and the growth of the
marshes along the present coast of the Wash may still be
observed. Occasional incursions of the sea over these
low-lying tracts have taken place in historical times, but
much land has been reclaimed by artificial banks in the
course of the last few centuries. The Fenland is indeed

[1] Miller and Skertchly, *The Fenland* (1878), p. 5.

the Holland of England, for that country is but the
delta of the Rhine: the surface features of the two areas
are strikingly similar, as might be expected from their
similar origin and history.

The rivers of Cambridgeshire belong to one main
drainage-system—that of the Cam and Ouse. The
principal waterway runs in a general N.N.E. direction
through the county along the median valley, and all the
lateral streams ultimately find their way into it. This
Rhee-Cam-Ouse river occupies a longitudinal valley running along the strike of the beds and following very nearly
the main outcrop of the Chalk Marl to a point as far north
as Waterbeach. Thus it forms a 'subsequent[1]' river, for
it does not flow down or against the dip-slope of the strata
but along the line of strike.

The Rhee rises at Ashwell in Hertfordshire. The
springs which give rise to it are thrown out by the Chalk
Marl. It flows along the longitudinal valley in the Gault
in a general north-easterly direction, and at Hauxton is
joined by the united waters of the Granta and Lin rivers,
forming the river Cam. The branch of the Cam known
as the Granta rises near Quendon in Essex and flows
across the strike of the Cretaceous beds and against their
dip. Streams with such a course have been called
'obsequent[2]' streams and must be later in date than
the formation of the escarpment through which they cut.
The Granta in fact was once a short stream which ran
westwards from the face of the escarpment into the
longitudinal valley. It has cut its way back through
the whole Chalk plateau to its present source at Quendon,
and, since the existence of this 'obsequent' valley has

[1] W. M. Davis, *Geogr. Journ.* vol. v. (1895), p. 127.
[2] W. M. Davis, *loc. cit.*

been proved to be even in Essex previous to the deposition of the Chalky Boulder Clay (see p. 161), we can form some idea of the great antiquity of the drainage system now represented by the Cam.

The valley of the Lin is also occupied by an 'obsequent' stream which flowed along the same course before glacial times (see p. 161).

The Granta flows from Quendon, past Newport, Wenden, Chesterford, and Whittlesford to Shelford, where it is joined by the Lin which also rises on the Chalk and makes its way past Linton, Abington, and Babraham to Shelford.

The united rivers then flow west-north-west to Hauxton, there to join the Rhee.

From this point the river is known as the Cam and flows as a 'subsequent' stream along its north-easterly longitudinal valley, through the town of Cambridge, past Fen Ditton, Horningsea, Waterbeach, and Upware to Thetford, where it joins the Ouse.

The most important tributary of the Cam from the west is the river Bourn, which rises on the high ground near the village of Bourn as the overflow of the water in the Lower Greensand and finds its way over a low place in the rim of Gault bounding it on the east. It flows past Bourn in an easterly direction to join the Cam just below Grantchester. The Bourn is what is known as a 'consequent' stream because it flows down the dip-slope of the beds. The ridge of Chalk Marl that stretches with considerable interruptions between Madingley, Coton, and Castle Hill, forms a low watershed, the drainage from its north side falling into the Ouse while that from its south side finds its way to the Cam.

There is an 'obsequent' stream on the east which rises

near Fulbourn, but becomes a 'subsequent' stream east of
Teversham owing to abstraction. Its ancient course which
is marked by patches of gravel would bring it directly
westward to Fen Ditton. The 'subsequent' stream is in
its turn captured by a small 'obsequent' stream near
Swaffham which flows westward into the Cam. From
Little Wilbraham another 'obsequent' stream flows N.W.
to the Cam ; and after the junction of the Ouse and Cam
at Thetford we meet with a succession of 'obsequent' rivers
draining the high ground in the east and running against
the dip of the beds to fall into the Ouse. Such are the
Ousel, Ivel, Lark, Little Ouse, and Stoke.

The river Ouse itself between Bluntisham and Thetford
is a true 'consequent' stream, and is abstracted at Thetford by the 'subsequent' river occupying the longitudinal
valley.

According to Prof. W. M. Davis the elaboration
of the rivers is the result of two cycles of denudation.
At the close of the first cycle, when the land had been
reduced to a 'peneplain' or lowland of faint relief, an
uplift took place which started the second cycle of
denudation by renewing the activities of the rivers, and
this has led to the high degree of perfection in the
adjustment of the streams to structures. This drainage-
system was however determined in pre-glacial times,
and after the departure of the ice the land was left
swathed in a covering of clay of variable thickness and
possibly of somewhat irregular and patchy distribution.
The country however stood then and stands now at a lower
level than it did before the incoming of glacial conditions,
for the buried channels filled with Boulder Clay, silt, etc.
are sometimes below the present sea-level (see p. 156). The
water running off the clay-covered surface which the ice

had left flowed mainly into its ancient channels or at any
rate along the lines of their depressions, and the general
coincidence of direction in the pre- and post-glacial stream-
courses shows that the mantle of drift was not sufficiently
thick to obliterate the main features of the landscape.
Thus the principal rivers are the revived successors of the
pre-glacial streams, though in a minor degree they have
been "superimposed by sedimentation[1]."

The great antiquity of our river-system is shown by
the prevalence of 'subsequent' streams, for all young
river-systems have their 'consequent' streams more
strongly developed. In pre-glacial times the 'consequent' streams had all been 'beheaded' owing to their
upper portions being diverted by the strong 'subsequent'
streams; and the 'obsequent' rivers had likewise cut
back long and important valleys.

Since the glacial period subaerial denudation has
proceeded apace, for we find that the steepest slope of the
chalk scarp over which the Boulder Clay was spread is
distant 1¼ miles west of the present face of the escarpment (see p. 159).

In the northern region of Cambridgeshire the flat
Fenland is only at a very slight altitude above the sea.
At Upware, for instance, the level of the Cam is only
about 12 feet above low-water level at Lynn Deeps, so
that artificial drainage by dykes is general. A few
sluggish meandering streams traverse the plain to empty
themselves into the Wash, amongst which may be mentioned the Old Croft river which marks the former course
of the Great Ouse before it was artificially diverted into
its present northerly direction.

[1] G. K. Gilbert, "Geology of the Henry Mountains," *U. S. Geol. Surv.*
(1877), p. 143.

THE JURASSIC BEDS.

General remarks. The members of the Jurassic System which are found in this area occupy only its northern and north-western portion, with the exception of the Upware inlier. In the immediately adjoining counties of Bedford and Huntingdon there are, however, several sections of interest which are readily accessible; and in order to give a clear idea of the relations, etc. of the beds it will be necessary briefly to refer to them.

The only portions of the Jurassic System which occur in Cambridgeshire belong to the Middle and Upper Oolites, from the Oxford Clay to the Kimeridge Clay inclusive.

One of the features of the series in this part of England is the absence of the typical calcareous development of the Corallian to which we are accustomed in the south-western counties and in Yorkshire. In Bedfordshire, Cambridgeshire, and Lincolnshire there is an almost unbroken succession of clays from the Oxford Clay to the Kimeridge Clay. Some subsidiary and thin limestone bands are indeed intercalated, and the Upware Corallian limestone reef has long been famous, but practically the Jurassic deposits of this region consist of argillaceous material. It used even to be held by some authors that with the exception of the Upware limestone the Corallian beds were here entirely absent. But Prof. Seeley[1] in 1861 showed that they were palæontologically represented by part of the clay formation which had been included in the Oxford Clay; and more recently Mr T. Roberts[2] has brought forward considerable evidence in favour of this view not only for Cambridgeshire but also for Lincolnshire[3].

The Oxford and Kimeridge Clays which are fairly persistent in their characters from Dorsetshire to Yorkshire, show no peculiar lithological features in this district.

The Portland and Purbeck beds are wholly wanting and perhaps never existed in this part of England.

The general succession of the Jurassic rocks of Cambridgeshire is as follows:—

Upper Oolites { Kimeridge Clay.

Middle Oolites { Corallian { Ampthill Clay = { Coral Rag. { Coralline Oolite. { Lower Calcareous Grit. Oxford Clay.

THE OXFORD CLAY.

General remarks. The lowermost beds of the Oxford Clay do not occur in the Cambridge district but are met with near Peterborough and in Northamptonshire where the Kellaways Rock is quarried[4].

The western limit of the Oxford Clay runs in an irregular line from Bedford to Peterborough, being marked

[1] *Geologist*, vol. IV. (1861), p. 552.

[2] T. Roberts, *The Jurassic Rocks of Cambridge* (Sedgwick Essay, 1886).

[3] T. Roberts, *Q. J. G. S.* vol. XLV. (1889), p. 545.

[4] T. Roberts, *Jur. Rocks of Cambs.* p. 10. Judd, *Geol. of Rutland* (*Mem. Geol. Surv.*), 1875, p. 232.

by the outcrop of the Lower Oolites of east Northamptonshire. It forms the broad valley of the Ouse in the
neighbourhood of Bedford and the low-lying flat country
or plain to the north, as may be well seen from the Lower
Greensand escarpment at Sandy.

Its whole outcrop in Cambridgeshire is considerably
obscured by drift, but its eastern limit must lie somewhere
to the east of the River Ouse, between Bedford and
Huntingdon.

The unconformable junction of the Cretaceous with the
Jurassic beds is well exposed in the railway cutting west
of Sandy Station, where the Lower Greensand is seen
resting on the denuded edges of the Oxford Clay.

Characters and thickness. The Oxford Clay
consists of dark blue or bluish grey tenacious clays with
thin bands of sandy argillaceous limestones or lines of
septaria. The clays are occasionally crowded with crystals
of selenite (sulphate of lime), and nodules of pyrites
(sulphide of iron) are not uncommon.

The Oxford Clay has not been pierced in the Cambridge
district. At Bluntisham a well 300 feet deep did not
reach its base, but its thickness is not as much as 700 feet,
as stated in the Survey Memoir[1], for most of the Ampthill
Clay is there included with the Oxford Clay. According
to Mr T. Roberts the thickness is probably about 500 feet.

Occurrences and exposures. Near Sandy in Bedfordshire there is a good exposure of Oxford Clay with
septarian nodules and thin calcareous bands in the brick
pits at the foot of the Lower Greensand cliff on the north
side of the Great Northern Railway. Fossils are fairly

---

[1] *Mem. Geol. Surv. Explan. Quart. Sheet* 51, S.W. p. 5.

plentiful here, and from them we should infer that the beds belong to the upper part of the Oxford Clay.

The lowest beds seen in the neighbourhood are those dug in the brick-pits near St Neots, and belong to the *Am. ornatus* zone of Prof. Judd[1]. There is a greyish sandy limestone in these pits which Prof. Seeley called the "St Neots' Rock[2]," but Mr Roberts does not appear to attach much importance to it either palæontologically or stratigraphically.

At Godmanchester a brickyard exposes Oxford Clay of probably a somewhat lower horizon than that of St Ives. A thin sandy limestone is found here also near the top of the clay.

West of St Ives there is a large brickyard where the best and most accessible section of the Oxford Clay near Cambridge is to be seen. The "St Ives Rock" (see p. 16), supposed to be of the age of the Lower Calcareous Grit, caps the clays in the pit, and these must therefore be referred to the very top of the Oxford Clay, *i.e.* to the *Am. cordatus* zone.

The succession in the section (Fig. 2, p. 17) is as follows[3]:—

		ft.	in.
(1)	St Ives Rock—a brown ferruginous limestone	3	0
(2)	Dark blue clay	8	0
(3)	Layer of calcareous nodules	0	9
(4)	Dark blue clay with clusters of selenite crystals	15	0
(5)	Layer of calcareous nodules (irregular) resting on thin layer of clay	0	9—10

[1] Judd, *Geol. of Rutland* (*Mem. Geol. Surv.*), 1875, p. 232.

[2] *Ann. Mag. Nat. Hist.* ser. 3, vol. x. (1862), p. 105. See also T. Roberts, *op. cit.* p. 13.

[3] T. Roberts, *op. cit.* p. 16.

		ft.	in.
(6)	Greyish argillaceous limestone	1	0
(7)	Blue clay worked down to	16	0

Fossils—particularly *Gryphœa dilatata*—are fairly abundant in the clays.

On the railway line from St Ives to Bluntisham and Sutton calcareous bands of rock are seen in the clay and by their dip prove the existence of a low anticlinal below the valley of the Ouse[1].

Palæontology. The following fossils have been recorded from the Oxford Clay of this district.

	Sandy	St Neots	St Ives
REPTILIA.			
Cimoliosaurus plicatus Phil.			
„ *Richardsoni* Lyd. ...			
Ichthyosaurus thyreospondylus Owen ...			
Ophthalmosaurus sp.	×		
Peloneustes philarchus Lyd.			
Pliosaurus Evansi Seeley		×	
„ *ferox* Sauvage			
MOLLUSCA.			
Cephalopoda.			
Ammonites Achilles D'Orb. (Godman-			
chester)			×
„ *athletus* Phil.		×	×
„ *babeanus* D'Orb.			×
„ *Bakeriœ* Sow.(Godmanchester)			×
„ *cordatus* Sow.(Godmanchester)	×		×
„ *crenatus* Brug.			×
„ *Duncani* Sow.		×	
„ *Eugenii* Rasp.		×	×
„ *excavatus* Sow.			×
„ *flexuosus* Munst.			×
„ *Goliathus* D'Orb.			×
„ *hecticus* Rein.			×

[1] *Mem. Geol. Surv. Explan. Quart. Sheet* 51, S.W. p. 6.

	Sandy	St Neots	St Ives
MOLLUSCA (*cont.*).			
Cephalopoda (*cont.*).			
Ammonites Jason Rein.		×	×
„ *lophotus* Ziet.			×
„ *mariæ* D'Orb.	×		×
„ *oculatus* Phil.(Godmanchester)			×
„ *perarmatus* Sow.			×
„ *rupellensis* D'Orb.			×
„ *trifidus* Sow.	×		×
„ n. sp.			×
Belemnites abbreviatus Mill.			×
„ *hastatus* Montf.	×	×	×
„ *obeliscus*	×		
„ *puzosianus* D'Orb. (Godman-chester)	×	×	×
„ *puzosianus* var. *verrucosus* Phil.	×		
Belemnoteuthis sp.			×
Nautilus calloviensis Oppel.			×
Gasteropoda.			
Alaria trifida Phil.	×		×
Cerithium Damonis Lyc.			×
„ *muricatum* Sow.		×	
„ sp.			×
Lamellibranchiata.			
Astarte robusta Lyc.		×	
„ sp.			×
Avicula inæquivalvis Sow.	×		×
Cardium Crawfordi Leck.			×
Cucullæa concinna Phil.		×	×
Exogyra nana Sow.			×
Gryphæa dilatata Sow. (Godmanchester)	×	×	×
Isocardia sp.			×
Leda lachryma Quenst.			×
„ sp.			×
Lima rigida Sow.			×
Modiola bipartita Sow.	×		×
Nucula elliptica Phil.			×
„ *nuda* Phil.			×
„ *ornata* Quenst.	?×		×
„ *turgida* Bean			×
Ostrea gregaria Sow.			×
Pecten fibrosus Sow.		×	
Perna sp.	×		×
Pholadomya Phillipsi Morr.			×

	Sandy	St Neots	St Ives
MOLLUSCA (*cont.*).			
Lamellibranchiata (*cont.*).			
Pinna mitis Phil. ...			×
Thracia depressa Sow. ...			×
Trigonia clavellata Sow. ...			×
„ *costata* Sow. ...	×		
„ *elongata* Sow. ...			×
„ *Pellati* Mun. Chal. ...			×
BRACHIOPODA.			
Rhynchonella lævirostris McCoy ...			×
„ *varians* Schloth. ...		×	×
Terebratula oxoniensis (Walker M.S.) Dav. ...			×
Waldheimia impressa Von Buch ...			×
			×
ANNELIDA.			×
Serpula sp. ...			×
Vermilia sulcata Sow. ...			×
CRUSTACEA.			
Eryma Babeaui Etallon ...			×
„ *Georgei* Carter ...			×
„ *Mandelslohi* Meyer ...			×
„ *? pulchella* Carter ...			×
„ *ventrosa* Meyer ...			×
„ *Villersi* Morière ...			×
Eryon sublævis Carter ...			×
Glyphea hispida Carter ...			×
„ *regleyana* Meyer ...			×
Goniocheirus cristatus Carter ...			×
Magila dissimilis Carter ...			×
„ *levimana* Carter ...			×
„ *Pichleri* Oppel. ...			×
Mecocheirus socialis Meyer ...			×
Pagurus sp. ...			×
Pseudastacus sp. ...			×
ECHINODERMATA.			
Acrosalenia sp. ...			×
Pentacrinus sp. ...			×

THE LOWER CALCAREOUS GRIT.

General remarks. Two names have been given to this bed in our area from the localities in which it has been detected, namely Elsworth and St Ives. Prof. Seeley, who first described these limestones[1], held that they were on different horizons, the Elsworth Rock being in the highest zone of the Oxford Clay and the St Ives Rock 130 feet below it. But the palæontological evidence is strongly in favour of the two beds being identical in age though they differ slightly in lithological characters. It would seem that between St Ives and Elsworth the beds are bent into a broad anticlinal, the axis of which lies not far south of the St Ives brickyard and the River Ouse[2].

In order to present the evidence more clearly the two Rocks are here described separately.

(a) *The Elsworth Rock.*

In the village of Elsworth, situated about 8 miles W.N.W. of Cambridge, there are exposures of this rock on the banks of the stream which flows through the hamlet. In the immediate vicinity the rock is also seen by the roadside and in pits. It consists of two bands of limestone separated by a brownish-black clay. Prof. Seeley had a pit sunk into it, and describes the lower limestone as[3] "a dark-blue homogeneous limestone, which I can compare to nothing but the unseptarious cement-stones of the clays. The oolitic grains were abundant and as deeply ferruginous as though they had been exposed to the air, while, scattered

---

[1] *Ann. Mag. Nat. Hist.* ser. 3, vol. x. (1862), p. 98 *et seq.*

[2] T. Roberts, *op. cit.* p. 19 *et seq.*

[3] Seeley, *Ann. Mag. Nat. Hist.* 3, x. (1862), p. 99; and Bonney, *Camb. Geol.* p. 12.

irregularly about, branching and interlacing, were masses of undecomposed iron pyrites." The limestone measured from 3 to 7 feet in thickness, and this variation was ascribed to contemporaneous denudation. The overlying clay was 5 feet thick and contained a small variety of *Ostrea Marshi.* The upper limestone was only 18 inches thick but much resembled the lower bed except in being rather more sandy and in places less oolitic.

Near Bluntisham a rock similar in lithological characters has been met with, and again in a well near Bourn.

The dip of the Elsworth Rock is southerly. The fossils recorded from it come from the lower limestone band (see list on p. 20).

(b)   The St Ives Rock[1].

This Rock consists of 3 feet of ferruginous limestone and calcareous clay, and caps the Oxford Clay in the St Ives brickyard (Fig. 2). The minute subdivisions of this Rock are as follows[2]:—

	ft.	in.
(a) Yellowish-brown calcareous clay, thin and irregularly bedded. Fossils not abundant	0	7
(b) Two thin beds of brown ferruginous limestone, the uppermost containing decomposed nodules of iron pyrites. Fossils : *Exogyra nana, Goniomya literata,* etc.	1	6
(c) Brown sandy fossiliferous limestone containing oolitic grains of oxide of iron, like those in the Elsworth Rock. It is very nodular towards the base and here the fossils are most abundant ... ... -	0	6

[1] Seeley, *Ann. Mag. Nat. Hist.* ser. 3, vol. x. (1862), p. 101. Bonney, *Camb. Geol.* p. 11.   T. Roberts, *op. cit.* p. 21.

[2] T. Roberts, *op. cit.* p. 22.

ft. in.

(*d*) Grey calcareous sandy clay, brown in its upper part.
Below it passes down into the Oxford Clay. It contains
Pinna, Myacites recurva, Gryphœa dilatata, Pleurotomaria,
etc. 0 5

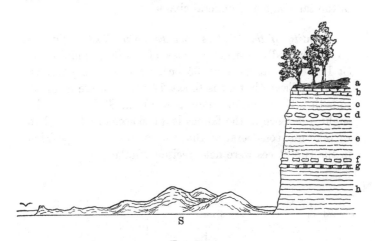

FIG. 2.

		ft.	in.
⤳	Alluvium		
a.	Thin clay (Ampthill Clay)		
b.	St Ives Rock	3	0
c.	Dark blue clay	8	0
d.	Thin layer of calcareous nodules	0	9
e.	Dark blue clay with aggregated crystals of selenite	15	0
f.	Irregular layer of calcareous nodules	0	9—10
	(separated from *g* by thin layer of clay)		
g.	Thin bed of greyish argillaceous limestone ...	1	0
h.	Blue clay		
S.	Spoil heaps		

The lower part of the St Ives Rock is very like that
of Elsworth and the faunas in the two (see below and list)
are very similar.

A rock believed to be a continuation of that of St Ives

has been pierced in well-borings at Swavesey and near Stretham Ferry. There are also fossils in the Wood-wardian Museum labelled from Holywell which by their species and the matrix show that they come from a rock of the same age and general character.

Identity of the St Ives and Elsworth Rocks. Of the 45 undoubtedly distinct species of fossils found in the St Ives Rock at St Ives, 35 occur also in the Elsworth Rock of Elsworth, that is to say 77 per cent. are common to both. Echinoids are more plentiful at St Ives, but the slight difference in the faunas is quite accounted for when we take into consideration that the conditions of deposition in the two places were not precisely similar.

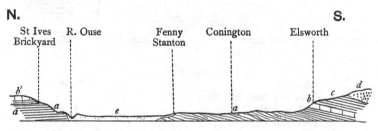

N.   **S.**

St Ives   R. Ouse   Fenny   Conington   Elsworth
Brickyard   Stanton

Fig. 3.   Section from St Ives to Elsworth.   (*T. Roberts.*)

a. Oxford Clay.   b′. St Ives Rock.   d. Lower Greensand.
b. Elsworth Rock.   c. Ampthill Clay.   e. Alluvium.

Again, the mutual resemblance of the fauna of the two rocks to that of the Lower Calcareous Grit of other areas is much in favour of their identity.

The stratigraphical evidence points to the same con-clusion: for the easterly dip of the St Ives Rock at St Ives, the southerly dip of the rock at Elsworth, and its presumed northerly dip near Bluntisham, as inferred from

the dip of the clay beds in the railway cutting, all point
to the existence of an anticlinal which though of slight
curvature is sufficient to explain how these rocks are on
the same horizon (Fig. 3).

Lithologically the characters of the rocks are similar,
as Prof. Seeley himself states.

Palæontology. The degrees of relationship of the
faunas of these thin limestones to those of the Oxford
Clay, the Lower Calcareous Grit and other beds, are
considered in detail by Mr Roberts[1]. Suffice it to say
that 50 per cent. of the species from the Elsworth Rock
and 66 per cent. of those from the St Ives Rock occur in
the Lower Calcareous Grit of other areas.

If we remember also that the Lower Calcareous Grit
of Yorkshire is from 80 to 100 feet thick while the
Elsworth and St Ives Rocks are only from 3 to 7 feet
thick the resemblance of the faunas is seen to be extremely
close. Moreover several species which so far have only
been recorded from the Lower Calcareous Grit, namely
Waldheimia bucculenta Sow., *W. Hudlestoni* Walk., and
Millericrinus echinatus Schloth., are found in the Elsworth
and St Ives Rocks, and three others, which are specially
characteristic of the Lower Calcareous Grit (*Ammonites
perarmatus* Sow., *Modiola bipartita* Sow., *Collyrites
bicordata* Leske), also occur. The clay beds, also, immediately below the St Ives Rock at St Ives contain
Ammonites perarmatus, Am. cordatus and the other ammonites which mark the highest zone of the Oxford Clay.

[1] *op. cit.* p. 28.

LIST OF FOSSILS FROM THE ST IVES AND ELSWORTH ROCKS[1].

	Oxford Clay and older beds	Elsworth Rock	St Ives Rock	Holywell	Lower Calcareous Grit of other areas	Newer beds
PISCES.						
Hybodus grossiconus Ag. ...	×		×		×	
,, sp.		×				
MOLLUSCA.						
Cephalopoda.						
Ammonites Achilles D'Orb. ...		×			×	
,, *canaliculatus* Münst.		×			×	
,, *convolutus* Quenst. ...	×	×			×	
,, *cordatus* Sow. ...	×	×	×	×	×	×
,, *Goliathus* D'Orb. ...		×				×
,, *Henrici* D'Orb. ...		×	×	×		
,, *perarmatus* Sow. ...	×	×	×		×	
,, *planicordatus* Seeley M.S.		×	×	×		
,, *plicatilis* Sow. ...		×			×	×
,, *vertebralis* Sow. ...	×	×	×	×	×	×
Belemnites hastatus Montf.	×	×	×		×	
,, *Oweni* ? Pratt		×				
Nautilus perinflatus Foord and Crick	×	×				
Gasteropoda.						
Alaria bispinosa Phil.	×			×	×	×
Natica Calypso D'Orb. var. *tenuis* Hudl.		×	×	×	×	
,, *Clymenia* D'Orb. ...		×				×
Nerinea sp.		×				
,, sp.				×		
Patella cf. *mosensis* Buv. ...		×			×	
Pleurotomaria granulata Lyc. (non Sow.) ...	×	×	×	×	×	
,, *Münsteri* Roem.		×	×	×	×	×
,, sp.		×				
Turbo Meriani Goldf.		×	×	×		

[1] T. Roberts, *loc. cit.* pp. 25—27.

	Oxford Clay and older beds	Elsworth Rock	St Ives Rock	Holywell	Lower Calcareous Grit of other areas	Newer beds
MOLLUSCA (cont.).						
Lamellibranchiata.						
Anatina undulata Sow....	×		×		×	×
Anomia sp.			×			
Arca æmula Phil.		×			×	×
„ sp. (with coarse ribs)		×				
„ 2 sp.			×			
„ terebrans Buv.		×				×
Astarte ovata Smith		×		?		×
„ robusta Lyc.	×	×	×			
„ sp.		×				
Avicula braamburiensis Sow.	×	×			×	
„ expansa Phil.	×	×	×	×	×	×
„ inæquivalvis Sow.	×	×				×
„ ovalis Phil.	?	×			×	×
„ pteropernoides Blake and Hudl.		×				×
Cardium Crawfordi Leck.	×	×	×	×		
Cucullæa clathrata Leck.	×	×				
„ elongata Phil.		×				×
„ oblonga Sow.	×	×				×
Exogyra nana Sow.	×	×	×		×	×
Goniomya literata Sow....	×	×	×	×	×	×
Gryphæa dilatata Sow....	×	×	×		×	×
Hinnites abjectus Phil.	×	×				
„ Sedgwicki Seeley M.S.		×				
„ velatus Goldf.		×				×
Isocardia globosa Bean		×	×	×		
Lima duplicata Sow.	×	×				×
„ leviuscula Sow.		×	×		×	×
„ n. sp.		×				
„ pectiniformis Schloth.	×	×			×	
„ rigida Sow.	×	×	×			×
Lithodomus sp.		×	×			
Lucina ampliata ? Phil.			×	×		×
„ globosa Buv.	×	×				×
Modiola bipartita Sow....	×	×	×	×	×	×
„ cancellata Roem.		×				×
Myacites Jurassi Brong....	×	×	×	×	×	×
„ recurva Phil.	×	×	×	×	×	×

	Oxford Clay and older beds	Elsworth Rock	St Ives Rock	Holywell	Lower Calcareous Grit of other areas	Newer beds
MOLLUSCA (*cont.*).						
Lamellibranchiata (*cont.*).						
Myoconcha cf. *Sæmanni* Dollf. ...		×	×	×		×
Nucula sp.		×	×			
Opis angulosa D'Orb.		×				×
Ostrea discoidea Seeley			×			×
„ *flabelloides* Lam.	×	×	×		×	×
„ *gregaria* Sow.		×	×		×	×
Pecten articulatus Schloth. ...		×	×	×		×
„ *lens* Sow....	×	×	×	×	×	×
„ *vagans* Sow.	×	×	×	×		×
Perna mytiloides Lam.		?	×		×	×
Pholadomya æqualis Sow. ...		×	×	×	×	×
„ *cingulata* ? Ag. ...		×				
„ *concentrica* Roem....		×			×	
„ *parcicosta* Ag. ...		×			×	×
Pinna lanceolata Sow.	×	×	×		×	×
„ *mitis* ? Phil.		×				
„ sp.		×				
Placunopsis sp.		×				
Plicatula fistulosa Lyc.	×	×				×
Thracia depressa Sow.	×	×	×	×	×	×
Trigonia elongata Sow.... ...	×	×	×			×
„ *Hudlestoni* Lyc. ...		×				×
„ *perlata* Ag.		×				×
Unicardium depressum Phil. ...	×	×				×
BRACHIOPODA.						
Terebratula insignis Schüb. ...		×				×
Waldheimia bucculenta Sow. ...		×	×		×	
„ *Hudlestoni* Walk....			×		×	
ANNELIDA.						
Serpula sp.		×	×	×		
CRUSTACEA.						
Glyphea sp.		×				
Goniocheirus cristatus Carter ...			×			

	Oxford Clay and older beds	Elsworth Rock	St Ives Rock	Holywell	Lower Calcareous Grit of other areas	Newer beds
ECHINODERMATA.						
Apiocrinus sp.		×				
Cidaris florigemma Phil. ...			×		×	×
,, *Smithi* Wright		×				
Collyrites bicordata Leske ...			×	×	×	×
Holectypus depressus Leske ...	×		×	×	×	×
Millericrinus echinatus Schloth.		×			×	
Pentacrinus sp.		×				
Pseudodiadema versipora Woodw.		×	×		×	×
ACTINOZOA.						
Thecosmilia sp.		×				

THE AMPTHILL CLAY[1].

General remarks. Resting on the Lower Calcareous Grit there is in Cambridgeshire a clay with sufficiently distinct characters to separate it from the underlying and overlying beds. Sedgwick recognised its presence and called it the Tetworth Clay in his sections through the Cambridge district published with his lecture "On the Strata near Cambridge, etc." in 1861. It has been termed the Ampthill Clay from the village in Bedfordshire where it is well exposed. It has also been called the Bluntisham,

---

[1] H. G. Seeley, *Geologist*, 1861, p. 558, *Ann. Mag. Nat. Hist.* 3rd ser. vol. x. p. 101; Bonney, *Camb. Geol.* p. 15; T. Roberts, *op. cit.* pp. 35—50; *Q. J. G. S.* vol. xlv. (1889), p. 545; *Mem. Geol. Surv. Explan. Quart. Sheet* 51, N.E. p. 7 and *Explan. Sheet* 65, p. 6.

Tetworth and Gamlingay Clay. Much of it in our area
was included by the Survey with the Oxford Clay and the
boundary between the two is not well defined owing to
the scarcity of exposures and the covering of drift. The
upper limit of the Ampthill Clay is drawn at the base of
a phosphatic nodule bed above which lies true Kimeridge
Clay.

Characters and thickness. The Ampthill Clay
consists mainly of black tenacious clays, sometimes carbonaceous, and weathering to a greyish-blue colour.
Crystals of selenite and some pyrites and a few phosphatic nodules occur in the beds, and thin limestone
bands, occasionally interrupted or nodular, are also found
at various horizons. The fossils are often fragmentary,
though plentiful, and are never pyritised, and by this
character Mr Roberts says the clay is easily distinguished
from the Oxford Clay. The thickness of the Ampthill
Clay is a matter of some uncertainty, but at Over according to the Survey Memoir[1] 200 feet of 'blue clay' was
pierced, all of which according to Mr Roberts[2] belongs
to the Ampthill Clay. At Cottenham the Clay is thinner and at Chettering Farm, $2\frac{1}{2}$ miles N.W. of Upware, it
is only 26 feet thick. Thus it appears to "thin out
rapidly in an easterly direction and may possibly dis-
appear altogether before Upware is reached, where it is
replaced by the Corallian limestones." (See p. 29.)

Occurrence and exposures. The lower limit or
western boundary of the Ampthill Clay is ill defined
owing to the covering of drift and lack of exposures.
The Clay is worked at Everton, west of Great Gransden,

---

[1] *Mem. Geol. Surv. Explan. Quart. Sheet* 51, S.W. p. 163.
[2] T. Roberts, *op. cit.* p. 48.

and north of Caxton; it occurs east of St Ives and has
been recognised to the north of Needingworth, in the
railway cutting west of Bluntisham, and at Fenton three
miles north-west of Somersham. So that its western
boundary must lie to the west of these places. In the
southern part of our area the outcrop of the Lower
Greensand forms its boundary on the east at Everton,
Great Gransden, Eltisley, and Papworth.

This narrow band of Ampthill Clay widens out north
of Papworth and forms the low ground north of Boxworth,
around Longstanton, Over, and Willingham, being bounded
on the east by the Lower Kimeridge Clay of Knapwell,
Oakington, and Willingham. Beyond Somersham it sinks
beneath the Fenland, but Mr Roberts has traced it
through Lincolnshire as a well defined band with the
same characters as in Cambridgeshire[1].

Of the exposures of the Ampthill Clay the best and
most accessible is in the pit at Gamlingay Bogs where
the following section is seen[2]:—

		feet.
(a)	Lower Greensand	10
(1)	Greyish-black clays	11
(2)	Grey argillaceous limestone	1
(3)	Black clays	9
(4)	Grey argillaceous limestone	1
(5)	Clay ... ... 2 to 3 feet	no longer seen.
(6)	Calcareous bed—not pierced	

There are two pits near Great Gransden, and in a
brickyard north of Boxworth good sections are exposed.
In the easterly pit at Boxworth occurs a limestone band
composed of two nodular beds and measuring two to three

[1] T. Roberts, "The Upper Jurassic Clays of Lincolnshire," *Q. J. G. S.*
vol. XLV. (1889), p. 545.

[2] T. Roberts, *Jur. Rocks of Camb.* p. 37.

feet in thickness. Prof. Seeley[1] describes a thin limestone in the Boxworth pit in the clay beds, which has been termed the "Boxworth Rock," but it appears to be very variable and of subordinate importance.

The long cutting on the Somersham and St Ives Railway near Bluntisham shows the Ampthill Clay, and two brick-pits at Fenton give exposures of it. At St Ives, immediately resting upon the St Ives Rock, there is a thin layer of clay which undoubtedly belongs to the horizon of the Ampthill Clay.

Economics. The Ampthill Clay is dug for brick-making purposes at Gamlingay, Boxworth, Fenton, and other places.

Agriculturally it has all the characters of the Oxford Clay.

Palæontology. The fauna of this Clay, taken with its stratigraphical position and relations, proves that the bed is on the horizon of the Corallian beds of other areas. There is indeed an admixture of Oxford Clay, Corallian, and Kimeridge Clay forms, but of the 41 species determined, 30 occur in the Corallian, 15 in the Oxford Clay and 19 in the Kimeridge Clay. The Oxford and Kimeridge Clay species are, moreover, principally those which in other areas pass up from the Oxford Clay into the Corallian or range from the Corallian into the Kimeridge Clay.

The only invertebrate peculiar to the Ampthill Clay is *Ostrea discoidea*. The saurians which are found in this bed are not recorded from any other formation.

[1] *Ann. Mag. Nat. Hist.* 3rd ser. vol. x. (1862), p. 104; *Mem. Geol. Surv. Explan. Quart. Sheet* 51, S.W. p. 7; T. Roberts, *op. cit.* p. 44.

LIST OF FOSSILS FROM THE AMPTHILL CLAY IN CAMBRIDGESHIRE[1].

	Oxford Clay	Corallian	Kimeridge Clay and Passage beds
REPTILIA.			
Cryptodraco eumerus Seeley			
Pliosaurus pachydeirus Seeley			
MOLLUSCA.			
Cephalopoda.			
Ammonites Achilles D'Orb.	×	×	
„ *alternans* Von Buch ...		×	×
„ *babeanus* D'Orb.	×		
„ *biplex* Sow.			×
„ *cordatus* Sow.	×	×	×
„ „ var. *cawtonensis* Bl. and H.		×	
„ *excavatus* Sow.	×	×	
„ *plicatilis* Sow.		×	
„ *serratus* Sow.		×	×
„ *vertebralis* Sow.	×	×	
Belemnites abbreviatus Mill.		×	×
„ *excentricus* Blainv.			
„ *nitidus* Dollf.		×	×
Nautilus hexagonus Sow.	×	×	?
Gasteropoda.			
Alaria bispinosa Phil.		×	
Cerithium muricatum Sow.		×	
Pleurotomaria sp.			
Lamellibranchiata.			
Arca longipunctata Blake		?	×
„ *rhomboidalis* Contej.			×
„ sp.			
Astarte supracorallina D'Orb.		×	×
Cardium sp.			
Corbula deshayesea Buv.		×	×
Cucullœa contracta Phil.		×	
Exogyra nana Sow.	×	×	×
Gryphœa dilatata Sow.	×	×	
Leda sp.			

	Oxford Clay	Corallian	Kimeridge Clay and Passage beds
MOLLUSCA (cont.).			
Lamellibranchiata (cont.).			
Lima pectiniformis Schloth.		×	×
Lucina aliena Phil.		×	
Myacites decurtata Phil.	×	×	
Nucula Menki Roem.		×	×
Ostrea deltoidea Sow.		×	×
„ *discoidea* Seeley			
„ *læviuscula* ? Sow.			
„ *Marshi* Sow.	×		
Pecten fibrosus Sow.	×	×	
„ *lens* Sow.	×	×	
„ *Thurmanni* Contej.			×
Plicatula sp.			
Thracia depressa Sow.	×	×	×
Trigonia clavellata Sow.	×	×	
„ *paucicosta* Lyc.	×		
BRACHIOPODA.			
Discina humphriesiana Sow.		×	×
ANNELIDA.			
Serpula tetragona Sow.		×	×
ECHINODERMATA.			
Cidaris Smithi Wright		×	×
Pentacrinus sp.			

THE CORALLIAN ROCKS OF UPWARE[1].

General remarks. The limestones of Upware form a most remarkable exception to the almost continuous series of clays which represent the Middle and Upper Oolites in the midland and eastern counties, for between the south of Yorkshire and Wheatley in Oxfordshire there is elsewhere practically no calcareous development of the Corallian beds.

The Upware limestones form a low broad ridge, rising about 20 feet above the surrounding plain and extending for about three miles north of the hamlet of Upware, in a direction nearly parallel to the course of the river. The beds are exposed in two quarries which are now more or less overgrown, and the sections described by former writers as showing the relation of the overlying beds to the Corallian limestones are unfortunately no longer visible. The Lower Greensand which forms the shelving sides of the ridge has been largely worked for its phosphatic nodules, but the workings are now abandoned.

On the western side of the ridge it was stated by Prof. W. Keeping that the limestones rose up as a bank against which the Kimeridge Clay rested unconformably, but this observation it has not been possible to verify; it was also stated that the base of this clay contained a

---

[1] Sedgwick, *Rep. Brit. Assoc.* 1845 (1846), *Trans. Sections*, p. 40; Fitton, *Trans. Geol. Soc.* ser. 2, vol. IV. (1836), p. 307; Bonney, *Camb. Geol.* (1875), p. 16 and Appendix I.; *id. Geol. Mag.* 1877, Dec. 2, vol. IV. p. 476; Blake and Hudleston, *Q. J. G. S.* vol. XXXIII. (1877), p. 313; *id. Geol. Mag.* 1878, Dec. 2, vol. V. p. 90; W. Keeping, *The Fossils, etc. of Upware* (Sedgwick Essay, 1883), p. 3 *et seq.*; *Mem. Geol. Surv. Explan. Quart. Sheet* 51, N.E. (1891), pp. 8—13; T. Roberts, *Jurassic Rocks of Cambridge* (Sedgwick Essay, 1886), pp. 51—60.

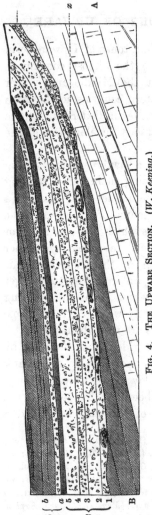

FIG. 4. THE UPWARE SECTION. (*W. Keeping.*)

(D) Gault.

    b. Unfossiliferous Gault—7 feet seen.   *a.* Phosphatic nodule bed, rich in fossils, about 5 inches.

Unconformity.

(C) Lower Greensand, about 12 feet.

    5. Clay bed—about 1 foot thick (has been referred to the Gault).

    4. Upper sand bed, or yellow 'silt.'

    3. Upper nodule bed.

    2. Lower sand bed, or yellowish 'silt.'

    1. Lower nodule bed and conglomerate.

    x = Junction bed of Lower Greensand, composed of Coral Rag fragments in a paste of Kimeridge Clay.

Unconformity.

(B) Kimeridge Clay.

(A) Coral Rag.

quantity of broken fragments of the limestones. Before
the deposition of the Lower Greensand beds there was
considerable denudation of the limestones and Kimeridge
Clay, and the Lower Greensand was banked up against
its sides and probably overlaid the summit, and the Lower
Gault was then deposited over it. Subsequent erosion
has again laid the limestone bare along the crest of the
ridge. The section on p. 30 (Fig. 4) shows the succession
and relations of the various formations. On the eastern
side the succession is much the same as on the western.

There has been considerable difference of opinion as to
the arrangement of the Corallian rocks in this ridge.
That they form an isolated lenticular patch in a great
mass of clay is agreed, for they have not been met with
in borings round the ridge. But Fitton[1] and Sedgwick[2]
regarded the limestones as rising in an anticlinal arch
through the newer beds, whereas Messrs Blake and
Hudleston[3] held that they formed a synclinal. A fault
between the two quarries has also been suggested to
explain the relations of the beds and the dips.

It has been proved[3] by the fossils that the creamy
white compact limestone of the south quarry belongs
to the Coral Rag and it is seen that these beds dip
north at a low angle. Below these hard limestone beds
there are now visible on the floor of this quarry beds
which contain the assemblage of fossils found in the
north quarry.

In the north quarry the limestones still dip north, but
are soft and oolitic or pisolitic in character and of quite a
different appearance from those in the sides of the southern

[1] *Trans. Geol. Soc.* ser. 2, vol. iv. (1836), pl. x. *a*, fig. 24.
[2] *Supplement* (1861), p. 22, section 1.
[3] Blake and Hudleston, *Q. J. G. S.* vol. xxxiii. (1877), p. 314.

pit. Messrs Blake and Hudleston have shown that the
fauna of these soft oolitic limestones belongs to the Coralline Oolite which underlies the Coral Rag of other areas.

In order therefore to explain how the northern pit,
which is practically at the same level as the southern pit,
shows a bed on a distinctly lower palæontological horizon
than the limestone of the southern pit but with a similar
northerly dip, we must suppose either that the beds rise
northwards from the floor of the southern pit into an
anticlinal but bend down again before reaching the
northern pit; or that a fault with a downthrow to the
south exists between the two quarries (Figs. 5 and 6).

Considering the fact that the Jurassic beds near
Bluntisham (see p. 18) are slightly bent over and that

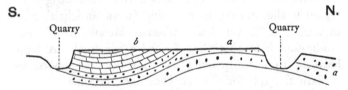

FIG. 5.   SECTION TO EXPLAIN THE STRUCTURE OF THE UPWARE RIDGE
BY MEANS OF FOLDS.   (*T. Roberts.*)

a.   Coralline Oolite.   b.   Coral Rag.

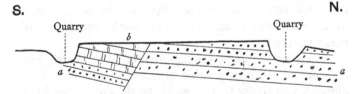

FIG. 6.   SECTION TO EXPLAIN THE STRUCTURE OF THE UPWARE RIDGE
BY MEANS OF A FAULT.   (*T. Roberts.*)

a.   Coralline Oolite.   b.   Coral Rag.

no faults have been proved to exist we may perhaps
regard the anticlinal hypothesis as the most probable
explanation.

Characters and thickness. It has been mentioned
that the limestones in the two quarries are dissimilar in
character and are referred to different horizons.

The Coral Rag.

In the southern quarry about 20 feet are exposed of a
creamy-white compact, rather irregular limestone, in parts
crystalline and with some layers composed entirely of
masses of coral (*Thamnastrea, Isastrea, Rhabdophyllia,*
etc.). Between these layers of coral the limestone is
usually softer, more oolitic, and earthy, and from these
softer beds most of the fossils have been obtained. The
fossils themselves are usually in the form of casts. *Litho-
domi* are not uncommonly found *in situ* in their borings
in the corals. *Cidaris florigemma,* the zone-fossil of the
Coral Rag of other areas, occurs in this limestone, and
Messrs Blake and Hudleston have no hesitation in corre-
lating it with the Coral Rag.

The Coralline Oolite.

In the northern pit (and in the floor of the southern
pit), on the other hand, there are found friable soft
calcareous beds composed of large or small oolitic grains
in a marly iron-stained matrix, with a few harder layers.
About 9 to 10 feet of these beds were seen in 1873[1].
The pit is now usually filled with water, but fossils can be

[1] *Mem. Geol. Surv. Explan. Quart. Sheet* 51, N.E. p. 9.

picked up at the water's edge. These beds Messrs Blake and Hudleston (*loc. cit.*) correlate with the Coralline Oolite of Yorkshire and other counties, which lies immediately below the Coral Rag, and is characterised by *Ammonites plicatilis.*

The limestones of the Upware ridge have never been pierced, so we have no knowledge of their base or of their true thickness, or of the underlying beds.

It has been proved by borings that they thin out rapidly on all sides, and it is generally held that this lenticular patch of limestone is of the nature of a modern coral-reef. In this way we may account for its isolated position and rich fauna, and in this connection the 'knoll-reefs' of Bolland (Yorks.) in the Lower Carboniferous, and the Triassic reefs described by Mojsisovics may be mentioned.

LIST OF FOSSILS FROM THE SOUTH QUARRY, UPWARE.

MOLLUSCA.

Cephalopoda.

Ammonites Achilles D'Orb.
,,   *mutabilis* Sow.
,,   *plicatilis* Sow.
,,   *vertebralis* Sow. var. *cawtonensis* Blake & Hudl.

Gasteropoda.

Alaria sp.
Amberleya princeps Roem.
Cerithium muricatum Sow.
Chemnitzia heddingtonensis Sow.
Emarginula Goldfussi Roem.
Fissurella corallensis Buv.
Littorina muricata Sow.
Natica Clymenia D'Orb.
,,   *Clytia* D'Orb.

MOLLUSCA (*cont.*).

Gasteropoda (*cont.*).

Neritopsis decussata Münst.

" *Guerrei* Héb. and Desl.

Pleurotomaria reticulata Sow.

" sp.

Pseudomelania striata Sow.

Trochotoma tornatilis Phil.

Trochus sp.

Lamellibranchiata.

Anomia suprajurensis Buv.

Arca æmula Phil.

" *anomala* Blake and Hudl.

" *contracta* Phil.

" *pectinata* Phil.

" *quadrisulcata* Sow.

Astarte aytonensis Lyc.

" *ovata* Smith

" sp.

Cardita ovalis Quenst.

Cardium cf. *delibatum* De Lor.

Cucullæa elongata Phil. (*non* Sow.)

Cypricardia glabra Blake and Hudl.

Exogyra nana Sow.

Gastrochæna moreana Buv.

Gervillia angustata Roem.

" *aviculoides* Sow.

Goniomya v. *scripta* Sow.

Hinnites velatus Goldf.

" cf. *corallina* Hudl.

Homomya tremula Buv.

Isoarca multistriata Etal.

" *texata* Münst.

Lima elliptica Whit.

" *gibbosa* Sow.

" *læviuscula* Sow.

" *rigida* Sow.

" *rudis* Sow.

MOLLUSCA (*cont.*).

Lamellibranchiata (*cont.*).

Lima sp.

Lucina globosa Buv.

  ,,   *moreana* Buv.

Modiola bipartita Sow.

  ,,   [*Lithodomus*] *inclusa* Phil.

  ,,   *subæquiplicata* Roem.

  ,,   sp.

Myacites decurtata Goldf.

  ,,   *recurva* Phil.

  ,,   *Voltzi* Ag.

Myoconcha Sœmanni Dollf.

  ,,   *texta* Buv.

Mytilus pectinatus Sow.

  ,,   *rauracicus* Greppin

  ,,   *ungulatus* Young and Bird

Opis arduennensis D'Orb.

  ,,   *corallina* Damon

  ,,   *paradoxa* Buv. ?

  ,,   *Phillipsi* Morr.

  ,,   *virdunensis* Buv.

Ostrea gregaria Sow.

  ,,   *solitaria* Sow.

Pecten articulatus Schloth.

  ,,   *inæquicostatus* Phil.

  ,,   *vimineus* Sow.

Perna subplana Etal.

Pholadomya decemcostata Roem.

Pinna lanceolata Sow.

Plicatula fistulosa Morr. and Lyc.

Quenstedtia lævigata Phil.

  ,,   ,,   var. *gibbosa* Hudl. ?

Trigonia meriana Ag.

BRACHIOPODA.

Rhynchonella sp.

Terebratula insignis Schübler var. *maltonensis* Oppel.

  ,,   sp.

CRUSTACEA.

Gastrosacus Wetzleri Meyer
Glyphea Münsteri Voltz. ?
Prosopon marginatum Meyer

ANNELIDA.

Serpula deplexa Phil. ?
„ *tetragona* Sow.
Vermicularia sp.

ECHINODERMATA.

Apiocrinus polycyphus Merian.
Cidaris florigemma Phil.
„ *Smithi* Wright
Hemicidaris intermedia Flem.
Millericrinus sp.
Pentacrinus sp.
Stomechinus gyratus Ag.

Collyrites bicordata Leske
Echinobrissus scutatus Lam.
Holectypus depressus Leske
Hyboclypus n. sp.
Pygaster umbrella Ag.

These are from the beds in the floor of the south quarry which are of the same age as the beds in the north pit, viz. the Coralline Oolite.

ACTINOZOA.

Isastræa explanata Goldf.
Montlivaltia dispar Phil.
Rhabdophyllia Phillipsi Edw. and Haime
Stylina tubulifera Phil.
Thamnastræa arachnoides Park.
„ *concinna* Goldf.

PORIFERA.

Scyphia sp.

LIST OF FOSSILS FROM THE NORTH QUARRY, UPWARE.

MOLLUSCA.

Cephalopoda.

Ammonites Achilles D'Orb.
    ,,   *perarmatus* Sow.
    ,,   *plicatilis* Sow.
    ,,   *trifidus* Sow.
Belemnites abbreviatus Mill.

Gasteropoda.

Littorina meriani Goldf.
    ,,   *muricata* Sow.
Pleurotomaria sp.

Lamellibranchiata.

Gervillia aviculoides Sow.
Isoarca texata Münst.
Modiola bipartita Sow.
Mytilus jurensis Mer.
    ,,   *ungulatus* Young and Bird
Opis Phillipsi Morris
Pecten fibrosus Sow.

ECHINODERMATA.

Collyrites bicordata Leske
Echinobrissus scutatus Lam. (very common in upper part
    of pit)
Holectypus depressus Leske (ditto)
Hyboclypus gibberulus Ag.
Pseudodiadema versipora Woodw.
Pygaster umbrella Ag.

THE KIMERIDGE CLAY.

General remarks. The Kimeridge Clay is the highest Jurassic bed now seen in the district. Its outcrop is comparatively small in southern Cambridgeshire, but in the northern part of the county it underlies a large tract of fenland.

It generally forms a low-lying plain more or less covered with alluvial deposits, but it also contributes to the formation of some of the islands in the Fens, especially the large one on which Ely stands.

Characters and thickness. The Kimeridge Clay consists of dark blue and blackish clays, often shaly and laminated, and occasionally so bituminous as to approach the character of the Kimeridge Coal. Thin layers of limestone in continuous bands or as lines of septarian nodules are also found. A seam of black phosphatic nodules marks the base of the clay. Crystals of selenite are scattered throughout its mass. The septarian nodules are flattened spheroidal or ellipsoidal masses and often of large size, commonly measuring as much as three feet across. The radiating cracks and sometimes the centre of these nodules are filled or lined with calcite in 'dogtooth' crystals with free inner terminations showing good crystalline faces. Fossils are fairly common throughout the clay.

The thickness of the clay is somewhat doubtful, but Mr T. Roberts[1] estimated it at over 142 feet. In Lincolnshire the formation is very much thicker according to Prof. Blake[2].

---

[1] T. Roberts, *op. cit.* p. 74; *Mem. Geol. Surv. Sheet* 51, N.E. p. 14.
[2] *Q. J. G. S.* vol. xxxi. (1875), p. 200.

Subdivisions[1].

Upper Division.   Clays and shales $\left\{\begin{array}{l}\text{(1) with } Discina\ latissima. \\ \text{(2) with } Exogyra\ virgula.\end{array}\right.$

Lower Division.   Clays $\left\{\begin{array}{l}\text{(1) with } Ammonites\ alternans. \\ \text{(2) with } Astarte\ supracorallina. \\ \text{(3) with } Ostrea\ deltoidea. \\ \text{Layer of phosphatic nodules.}\end{array}\right.$

These divisions and zones are mostly only of local value and differ from those in Lincolnshire[2], but both areas have the lower clay beds marked by an abundance of *Ostrea deltoidea*.

Occurrence. From Knapwell where the most southerly and westerly exposure in our area occurs a narrow band runs through Boxworth, Oakington and Cottenham, beyond which it spreads out beneath the fenland of North Cambridgeshire. The chief sections worthy of mention are those at Haddenham, Ely, and on the road to Littleport from Ely. At Haddenham half a mile west of the station there is a large brick-pit where the following section may be seen[3]:—

		ft.	in.
(a)	Soil	1	6
(1)	Ferruginous clay	2	0
(2)	Mottled clays with some calcareous nodules	4	0
(3)	Thin limestone beds, greyish in colour, but weathering white, very hard and compact and with cherty appearance...	1	3
(4)	Black tenacious clays with few fossils	2	0
(5)	Grey nodular limestone with *Ammonites mutabilis* and some small gasteropods	0	9
(6)	Black clay	4	0
(7)	Black clays crowded with very large specimens of *Ostrea deltoidea*, underlain by clay containing a considerable quantity of black phosphatic nodules...	5	0

[1] T. Roberts, *op. cit.* pp. 66—69; *Mem. Geol. Surv. Explan. Quart. Sheet* 51, N.E. p. 13.

[2] *Q.J.G.S.* vol. xlv. (1889), p. 559.   [3] T. Roberts, *op. cit.* p. 63.

Mr Roberts says, "The basement bed of the Kimeridge Clay, with its numerous phosphatic nodules, is fairly constant in this district, and with the clays crowded with *Ostrea deltoidea* overlying it, marks an easily recognised horizon. Similar beds occur at the base of the Kimeridge Clay near Oxford." The presence of this seam of phosphatic nodules may point to a slight denudation at the beginning of the Kimeridge Clay period or more probably to a temporary cessation of sedimentation. In the famous pit at Roslyn Hill, Ely, there is the best section of this clay in the neighbourhood[1]. Mr Roberts gives the following section seen on the northern side of the pit where the beds dip gently towards the west :—

		ft.	in.
(a)	Soil and reassorted clay	5	0
(1)	Bituminous papery shale with *Discina latissima*	7	0
(2)	Calcareous clay with interrupted lines of septarian nodules	0	7
(3)	Greyish black shale with some thin beds of more sandy clay ; near its base it is crowded with *Exogyra virgula*	3	0
(4)	Dark grey shales with *Ammonites alternans*, etc.	1	6
(5)	Thin ferruginous layer	0	2
(6)	Greyish shale with three sandy layers	5	0
(7)	Greyish sandy shale with *Trigonellites*	3	0
(8)	Sandy clay with interrupted lines of septarian nodules	1	0
(9)	Bluish clay with fragmentary fossils	1	6
(10)	Papery shale somewhat sandy and crowded with fossils, passing down into a more arenaceous bed. This is separated by a thin layer of clay from another sandy bed containing few fossils	0	6

[1] Seeley, *Geol. Mag.* vol. II. (1865), p. 529, vol. V. (1868), p. 347; Fisher, *Proc. Camb. Phil. Soc.* Part IV. (1867), p. 51, *Geol. Mag.* vol. V. (1868), pp. 408, 438 ; Bonney, *Camb. Geol.* p. 20 and Appendix II., *Proc. Camb. Phil. Soc.* 1872, Pt. XIII. p. 268; Blake, *Q. J. G. S.* vol. XXXI. (1875), p. 201; T. Roberts, *op. cit.* p. 66 ; *Mem. Geol. Surv. Explan. Quart. Sheet* 51, N.E. p. 15.

		ft.	in.
(11)	Bluish clay	1	0
(12)	Sandy shales with few fossils	1	0
(13)	Dark blue clays	9	0
(14)	Fissile sandy shales with *Astarte supracorallina* in abundance	0	9
(15)	Dark clay	3	0
		43	0

On the road from Ely to Littleport there are three brick-pits in all of which sections may be seen. In the pit situated at the fourth milestone from Ely near the summit of the high ground on which Littleport stands there is a good exposure of beds intermediate in stratigraphical position between those seen at Haddenham and those seen at Ely. The following section is given by Mr Roberts[1] :—

		ft.	in.
(a)	Soil and gravel		
(1)	Dark blue clays with small crystals of selenite and few fossils	6	0
(2)	Layer of isolated septarian nodules lying in greyish sandy and calcareous shale with *Astarte supracorallina* ...	0	8
(3)	Dark clays with some thin arenaceous beds ...	11	0
(4)	Grey very argillaceous limestone with *Trigonia*, etc.	1	0
(5)	Black clays	4	0
(6)	Very compact limestone, somewhat nodular in one part of the section	1	6
(7)	Dark blue tenacious clays	3	0

The beds here dip north-west at about 5°.

The papery shales with *Discina latissima* which are seen in the Roslyn pit are the highest beds of the Kimeridge Clay of this district, while the phosphatic nodule bed west of Haddenham forms its base.

---

[1] T. Roberts, *op. cit.* p. 71.

Economics. The Kimeridge Clay is dug for brick-making in many places, as at Haddenham, Ely, and Littleport. It is also used for the river banks and for tamping the dykes in the Fens.

Note. A limb-bone of a Plesiosaurian which was found between Littleport and Ely is marked with several grooves and scratches which point to the carnivorous habits of the larger fishes and reptiles[1].

LIST OF FOSSILS FROM THE KIMERIDGE CLAY OF
CAMBRIDGESHIRE.

REPTILIA.

Cimoliosaurus trochanterius Owen
           ,,   *truncatus* Owen
Dacosaurus maximus Plieninger
        ,,   sp.
Gigantosaurus megalonyx Seeley
Ichthyosaurus chalarodeirus Seeley
          ,,   *hygrodeirus* Seeley
          ,,   *trigonus* Owen
Metriorhynchus sp. ?
Peloneustes æqualis Phil.
        ,,   sp.
Pliosaurus brachydeirus Owen
        ,,   *brachyspondylus* Seeley
        ,,   *grandis* Owen
        ,,   *macromerus* Phil.
        ,,   *nitidus* Phil.
Steneosaurus sp.
Thalassemys Hughii Rütim.

PISCES.

Asteracanthus carinatus McCoy
        ,,   *ornatissimus* Ag.

[1] T. MᶜK. Hughes, *Trans. Vict. Instit.* May, 1889.

44   THE JURASSIC BEDS.

PISCES (*cont.*).

Ditaxiodus impar Owen
Eurycormus grandis Woodw.
Gyrodus sp.
„ *umbilicatus* Ag.
Hybodus acutus Ag.
„ *Fischeri* Newton
Ischyodus sp.
Lepidotus sp.
Leptacanthus semicostatus McCoy
Macropoma substriolatum Huxley
Pachycormus sp.
Sphenonchus sp.

MOLLUSCA.

Cephalopoda.

Ammonites alternans Von Buch
„ *biplex* Sow.
„ *calisto* D'Orb.
„ *cordatus* Sow. var. *excavatus* Sow.
„ *eudoxus* D'Orb.
„ *longispinus* Sow.
„ *mutabilis* Sow.
„ *rotundus* Sow.
„ *trifidus* Sow.
Belemnites abbreviatus Mill.
„ *explanatus* Phil. ?
„ *nitidus* Dollf.
Trigonellites (*Aptychus*) *latus* Park.

Gasteropoda.

Alaria trifida Phil.
Delphinula nassoides Buv.
Dentalium Quenstedti Blake

Lamellibranchiata.

Anomia Dollfussi Blake
Arca mosensis Buv.
„ *reticulata* Blake

MOLLUSCA (*cont.*).

Lamellibranchiata (*cont.*).

Arca rhomboidalis Contej.
„ *rustica* Contej.
Astarte ovata Sow.
„ *supracorallina* D'Orb.
Avicula œdilignensis Blake
„ *costata* Sow.
„ *dorsetensis* Blake
„ *echinata* Sow.
„ *inæquivalvis* Sow.
„ *nummulina* Blake
Cardium striatulum Sow.
Exogyra nana Sow.
„ *virgula* Defr.
Gryphœa dilatata Sow. ?
Lima pectiniformis Schloth.
Lucina minuscula Blake
Mactromya rugosa Roem.
Myacites sp.
Myoconcha sp.
Nucula Menki Roem.
„ *obliquata* Blake
Ostrea deltoidea Sow.
„ *gregaria* Sow.
„ *læviuscula* Sow.
„ *monsbeliardensis* Contej.
„ *solitaria* Sow.
Pecten demissus Phil.
„ *Grenieri* Contej.
„ *lens* Sow.
„ sp.
Perna Flambarti Dollf.
Pholadomya ovalis Sow.
Pleuromya donacina Ag.
Thracia depressa Sow.
Trigonia elongata Sow.
„ *Pellati* Mun. Chal.

BRACHIOPODA.

> *Discina latissima* Sow.
> „   sp.
> *Lingula ovalis* Sow.
> *Rhynchonella inconstans* Sow.
> „   *pinguis* Roem. ?
> *Terebratula Gesneri* Etal.

ANNELIDA.

> *Serpula intestinalis* Phil.
> „   *tetragona* Sow.
> *Vermicularia contorta* Blake
> *Vermilia sulcata* Sow.

ECHINODERMATA.

> *Cidaris spinosa* Ag.
> „   sp.

CRUSTACEA.

> *Cythere æqualis* Blake M.S.
> *Cytheridea Ruperti* Blake M.S.
> „   *triangulata* Blake M.S.
> *Pollicipes Hausmanni* Koch and Dun.

PROTOZOA.

FORAMINIFERA.

> *Cristellaria lævigata* D'Orb.
> *Marginulina gracilis* Corn.
> „   *lata* Corn.
> *Planularia strigilata* Reuss
> *Vaginulina harpa* Roem.
> „   *striata* D'Orb.

THE CRETACEOUS BEDS.

THE greater part of Cambridgeshire is occupied by beds belonging to this system, though concealed over large tracts by Post-Tertiary deposits. The Lower Greensand forms its basement bed. The Upper Chalk above the Zone of *Micraster cor-testudinarium* has been completely removed by denudation.

The "Cambridge Greensand" is of special interest, but the peculiar characters of the Lower Greensand, the change in the Gault when traced northward, and the rock-beds in the Chalk are all worthy of particular attention.

THE LOWER CRETACEOUS.

THE LOWER GREENSAND.

General remarks. The lowest beds of the Lower Greensand are entirely absent in England, and in our district only that part which is held to correspond to the very uppermost part of the Aptian of the Continent is present at all[1]. In the Midlands and the eastern part of England the so-called Lower Greensand represents only

[1] See *Text-book of Comparative Geology*, by Kayser and Lake, 1893, p. 293.

the higher divisions of the Lower Greensand of the south
and south-eastern counties. It is not believed to include
any part of the Neocomian of D'Orbigny. For this reason
it seems undesirable to introduce that term at all into
English Geology, especially in view of its different applications on the Continent and the many points of dispute
connected with it.

Relation to underlying and overlying beds. As
in the midland district, the Lower Greensand in the
eastern counties rests unconformably on the Upper Jurassic
rocks which prior to its deposition had suffered considerable erosion. Thus, while in Buckinghamshire it rests
sometimes on Purbeck and sometimes on Portland beds,
at Sandy in Bedfordshire it directly reposes on an eroded
surface of Oxford Clay, and as it is traced north-eastwards
by Gamlingay, Gransden, and Eltisley it appears gradually
to creep over the edges of higher and higher members of
the Upper Jurassic series. The Ampthill Clay lies immediately beneath it at Gamlingay and Caxton[1], and
further to the north-east at Cottenham and again at Ely
the Lower Greensand is seen to be underlaid by Kimeridge
Clay. At Upware it was recorded by Mr W. Keeping to
have been seen banked up against the Corallian limestone
knoll, while in the immediate neighbourhood to the west
and east of the ridge it rests on Kimeridge Clay. The
Upware section however is no longer exposed and the
original interpretation has been disputed.

This unconformable transgression of the Lower Greensand over the older Mesozoic beds is a notable feature in
the geological maps of the district.

[1] T. Roberts, *The Jurassic Rocks of Cambridgeshire* (Sedgwick Essay,
1886), p. 37.

In some places the Gault overlaps the Lower Greensand, or even entirely passes over its outcrop in such a
way as to suggest an unconformity; but at Gamlingay in
Cambridgeshire and West Dereham in Norfolk there
seems to be a complete passage between them.

The strike of the Lower Greensand in Cambridgeshire
and Bedfordshire is north-east and south-west. The local
distribution and character of the bed have probably
been influenced by the Palæozoic barrier or platform now
underlying Norfolk and Suffolk, part of which was probably
still above water at the time of its formation. In Norfolk
the strike bends round to a north and south direction.

Characters and thickness. The thickness and
characters of the Lower Greensand are very variable.
While in Buckinghamshire and Bedfordshire the beds of
yellow and white sand attain a thickness of over 200 feet,
e.g. 250 ft. at Fenny Stratford, and about 220 ft. at
Woburn, at Ampthill they are only 30 ft. thick because
much of the top has been removed by denudation.

At Sandy on the right bank of the Ivel the sands are
about 100 ft. thick, but throughout Cambridgeshire do not
exceed 70 ft. in thickness, frequently being very much
less, as for instance at Upware where the beds are only
12 ft. thick and seem on the point of thinning out altogether owing to the original unevenness of the surface of
deposition. At Wicken they are only about 8 ft. thick, but
in south-west Norfolk their thickness has increased and
averages 150 ft.

The composition and lithological characters of the
Lower Greensand likewise undergo a considerable amount
of modification when traced north-eastward along the
strike. In the southern part of its outcrop and throughout

Cambridgeshire the formation consists of brown, yellow, and white sands where exposed at the surface and oxidised, but where proved in borings they are of a deep green colour; they are more or less ferruginous, friable and soft, with some beds of loam, ironstone, gravel, and fine conglomerate intercalated. Near their base in Buckinghamshire there are pebbly beds containing brown phosphatic nodules —the so-called 'coprolites'—which have been worked at Brickhill, near Bletchley, and in Bedfordshire at Ampthill and around Sandy and Potton (see p. 53). At Woburn a bed of Fuller's Earth, 12 feet in thickness, occurs rather below the middle of the sandy series and has been dug for the last 150 years. The sands are frequently false-bedded and contain ironstone-concretions. Carstone, which is an indurated ferruginous sandstone, occurs in subordinate layers in Bedfordshire and in parts of Cambridgeshire, and is largely used for building stone. This hardening of the sand is however only a superinduced character of comparatively recent date and it does not mark any definite geological horizon. Seams of phosphatic nodules were found at Upware with pebbly sands and conglomerates, and again at Wicken. When traced further northwards, impersistent beds of clay are intercalated and alternate with sandy layers, but are specially characteristic of the middle part of the series, as at Heacham in Norfolk. A thick bed of carstone occurs at the top of the Lower Greensand in the latter county and is well seen on the foreshore at Hunstanton, while the lower sands (which resemble lithologically those of Sandy and Woburn) occupy a large area near Sandringham. From Heacham to Sandringham the Lower Greensand forms a marked escarpment.

At Hunstanton (Fig. 7, p. 84) the Lower Greensand

consists of a thick carstone with iron-sands and pebble beds
and a zone of nodules at the base of the series. These
contain derived and rolled fossils characteristic of the
Atherfield Clay as at Upware and Potton (*Ammonites
Martini, A. Deshayesi* and *Ancyloceras gigas*)[1]. The pebbles
scattered throughout the whole series and occurring in the
conglomerates also resemble very closely those of Potton
and Wicken, though they are generally of a smaller size.

The nodule beds of the Lower Greensand do not seem
to be all on the same horizon, and are impersistent
throughout the whole area. At Stretham phosphatic
nodules are distributed promiscuously throughout the
beds and are not aggregated into special layers.

Occurrence. In Buckinghamshire and Bedfordshire
the Lower Greensand has a wide surface-outcrop which
runs from Leighton Buzzard through Woburn, Ampthill,
Shefford, Sandy, and Potton. From Woburn to Shefford
it forms a range of hills, and at Sandy the river Ivel cuts
across the strike and its escarpment forms a picturesque
cliff 50 to 60 feet in height on the right bank. From
Sandy through Everton to Gamlingay in Cambridgeshire
the sands give rise to important features in the landscape,
and the difference of the scenery on the Lower Greensand
of this part from that of the rest of Cambridgeshire cannot
fail to arrest attention. The sandy steep-sided hills
scantily clothed with gorse, heather, broom, and Scotch fir,
and the wide breezy expanses of sterile common, supporting
on the poor soil little vegetation except gorse, are in
striking contrast to the tameness of the neighbouring

[1] Wiltshire, *Q. J. G. S.* vol. xxv. (1869), p. 188, and W. Keeping, *The
Fossils of Upware, etc.* (Sedg. Essay, 1879), pp. 33 and 56.

landscape, and remind one of the country around Hasle-
mere, at the western extremity of the Wealden anticline,
where the same formation gives rise to just the same type
of surface-feature but on a larger scale.

North-east of Potton the Lower Greensand is completely
concealed by a thick covering of Boulder Clay, from which
it emerges east of Boxworth with a narrow outcrop and
much reduced thickness. Tracing it further along its
strike in the same direction we find the outcrop widening
somewhat by the long straight track marked "Via Devana"
on the Ordnance Map, and stretching past Oakington to
Cottenham with a surface-breadth of about half a mile.
River-gravel then overlies it for some distance, but it
reappears near the Plough Inn on the Cambridge and Ely
road and passes under the fen to Upware and Wicken.
Further north on the left bank of the Ouse it forms a long
spur of higher ground reaching by Stretham, Wilburton,
and Haddenham to Aldreth. There is another patch—an
outlier—at Ely which forms the higher ground for about
$1\frac{1}{2}$ miles south-west of the Cathedral and to a less distance
to the north-east of it. A broad expanse of fenland then
conceals its northerly extension as far as Denver in
Norfolk.

Economics. The phosphatic nodules which occur
generally in layers in the Lower Greensand have been
worked in the neighbourhood of Cambridge at various
places along its whole outcrop. Chief among these localities are Brickhill, near Bletchley, Potton, Upware and
Wicken. (For analyses of the nodules see p. 61.)

As the phosphatic nodules are commonly associated
with pebble beds, the nodules have to be picked out,
and the refuse is commonly used for road metal, and

indeed these beds sometimes contain so little phosphate that they are useless for any other purpose than laying on the roads.

The Lower Greensand forms the most important water-bearing stratum of the district, and wells are sunk into it in many places[1]. Below the town of Cambridge the Lower Greensand is struck at a depth of about 125 feet; and at the Cherry Hinton Waterworks a soft brown sand belonging to it and containing water was reached at a depth of 176 feet from the surface.

The Lower Greensand does not possess in Cambridge-shire a sufficiently wide outcrop to be of much importance agriculturally, but where it forms the surface rock it may be traced across the fields by the redness and warmth of the soil. Around Sandy the soil on it is very poor and barren, but the wash from the sandy hills over the heavy clays of the surrounding plain has much to do with the fertility of the market-gardens of the Ivel valley.

The carstone is locally used for building purposes, and the sands are used for glass making.

Special localities.

Sandy and Potton[2].

Though these places are situated just outside the Cambridgeshire boundary yet they are so frequently visited from Cambridge on account of their accessibility and the existence at them of the only good exposures of

---

[1] See *Mem. Geol. Surv. Explan. Quart. Sheet* 51, S.W. p. 155 *et seq.*

[2] H. G. Seeley, "On the Potton Sands," *Proc. Camb. Phil. Soc.* I. (1866—7), p. 40; *Ann. Mag. Nat. Hist.* vol. xx. 1867, p. 23; J. J. H. Teall, *The Potton and Wicken Phosphatic Deposits* (Sedgwick Essay, 1875).

the Lower Greensand in the neighbourhood that an
account of them is necessary.

At Sandy, close to the railway station, there is a fine
cliff-exposure—over 50 ft. high—of the lower part of the
Lower Greensand of the district, consisting of yellow,
white, brown and red ferruginous ill-compacted false-
bedded sands. The upper beds are exposed in quarries
on Sandy Heath.

The loose lower sands by the railway contain numerous ironstone concretions, with their interior filled with
white sand. The mode of their formation according
to Jukes-Browne[1] is as follows:—The iron peroxide with
which the sands are charged separates into very thin
lines which simulate stratification. These thin layers
sometimes follow the lines of true-bedding and sometimes
those of false-bedding. After reaching a thickness of
about ⅛ inch and a distance apart of some 3 or 4 inches
they begin to bend towards each other at intervals
varying from 4 inches to a foot or more in length, and
ultimately meet, enclosing patches of sand. The ironstone
shell of each concretion thus formed increases in hardness
and thickness by the further segregation and elimination
of the iron from the surrounding sand and enclosed sandy
nucleus. Ultimately when the whole of the iron has
been concentrated in the shell the kernel of sand is left
soft and white. Another view that they are formed in
relation to joint planes by the segregation and infiltration
of the iron along these cracks seems to be more probable.

Layers of these ferruginous concretions in all stages of
formation and of various shapes—tabular, cubical, spherical
or irregular—may be seen in the pits, and when of a
fairly rectangular shape are known as 'boxstones.'

[1] *Mem. Geol. Surv. Explan. Quart. Sheet* 51, S.W. (1881), p. 12.

At the western end of the great cliff of sand at Sandy, immediately under a small bridge that crosses the line of rail to the brick-pits in the Oxford Clay, there is an excellent section in the bank showing the Lower Greensand resting unconformably on the eroded edges of the Oxford Clay, which here contains several thin bands of impure limestone (see p. 10).

The loose sandy beds, so well seen in this cliff at Sandy, forming the lower part of the Lower Greensand of this district, underlie the conglomerate which is worked on the top of the plateau nearer Potton for the nodules of phosphate of lime—wrongly called ' coprolites'—with which it is crowded.

The whole series varies much from place to place, so that very different sections may be seen within a few yards of each other. Teall[1] gives the following section on Sandy Heath and it may still be made out more or less plainly in the pits.

(1) Sands, slightly indurated, stratified horizontally ... ... 3 ft.

(2) Coarse ferruginous sand, false bedded, containing hard flaggy bands of deeply coloured sandstone ; the bands frequently ramifying in a peculiar
manner in consequence of the concretionary tendency
of the oxide of iron (limonite and hæmatite) ... ... 4—5 ft.

(3) Horizontally stratified sandstone, with small
pebbles ... ... 2 ft. 10 ins.

(4) Nodule bed ... ... 2 ft.

[(5) Sands, not exposed in this section.]

A well is stated[2] to have been sunk in these sands below the nodule bed for 50 ft.; and their thickness must be still greater.

[1] Teall, *op. cit.* p. 4.
[2] Brodie, *Geol. Mag.* vol. III. (1866), p. 153.

The nodule bed (No. 4 in this section) is composed of phosphatic nodules and pebbles, mixed in about equal proportions in a matrix of ferruginous sand, and, though commonly about 2 feet, sometimes reaches as much as 6 feet in thickness[1] or diminishes to 6 inches[2].

The Nodules. The nodules, which are of a light brown colour on the outside but have a much darker interior, consist of casts of molluscs, reptilian bones, etc. and of amorphous lumps of phosphatic material. Nearly all are in a more or less rolled and water-worn condition, and are frequently pierced by small boring *Modiolæ*. The nodules vary in size from that of a pea to a fowl's egg and sometimes are even larger.

It is now generally agreed that the great majority if not the whole of the phosphatised fossils are derivative. Most of the mollusca which can be identified amongst them belong to Jurassic species (see list, p. 59). The commonest fossil in this condition is *Ammonites biplex*, and Mr Teall says that at least half of the nodules are portions of casts of this ammonite.

Owing to the mixture of non-phosphatic pebbles in the conglomerate the nodules have to be picked out by hand.

The Pebbles. The pebbles themselves are of great interest and give some indication of the probable source of the sedimentary material. They are of various sizes, and consist of grey and red quartzite, white vein-quartz, lydianite, jasper, chert, ferruginous conglomerate and ironstone concretions. A block of gneiss of over 100 cubic inches in volume was found with other large boulders by Mr Teall.

[1] Bonney, *Camb. Geol.* (1875), p. 24.
[2] Teall, *op. cit.* p. 4.

The Fossils. In addition to the derived fossils which
are in a more or less imperfect and worn condition there
are some which occur in a totally different state of preservation. These are the ferruginous shells and boring
molluscs which represent the indigenous fauna that lived
contemporaneously with the formation of the bed. The
flora of the period is chiefly represented by fairly numerous
remains of coniferous trees.

These non-derivative fossils are generally not rolled
or water-worn. Their ferruginous condition is due to the
replacement of the calcite by iron oxide derived from the
surrounding sands by the agency of percolating water.

Of the species identified in this contemporary non-
derivative fauna eight are found in the Lower Greensand of the south of England, three occur at Faringdon,
and five of the others are peculiar to the Potton
deposit. Four or five of the fossils range up into higher
beds. The presence of rolled true Lower Greensand
fossils (*Ammonites Deshayesi*, etc.) in the nodule bed
proves that there were earlier Lower Greensand deposits
of the age of the Atherfield Clay and possibly belonging
to the lower part of the Hythe Beds in this or adjoining
areas which were broken up and eroded prior to the formation of the Potton bed. This fact taken in conjunction
with the character of the indigenous fauna and with the
evidence from Upware and from Kent and Surrey (see
p. 62) leads us to consider the Sandy and Potton deposits
to be approximately the equivalents of the upper part
of the Hythe Beds of the southern counties.

58   THE CRETACEOUS BEDS.

LIST OF NON-DERIVATIVE FOSSILS OF THE POTTON
LOWER GREENSAND.

BRACHIOPODA.

Rhynchonella antidichotoma Buv.
    „   *latissima* Sow.
Terebratula Dallasi Walk.
    „   *depressa* Sow.
    „   *prælonga* Sow.
Waldheimia tamarindus Sow.

MOLLUSCA.

Gasteropoda.
Pleurotomaria gigantea Sow.
Lamellibranchiata.
Cyprina Sedgwicki Walk. (? *C. angulata* Flem.)
Exogyra 2 sp.
Lucina vectensis Forbes ?
Modiola æqualis Sow.
Ostrea macroptera Sow.
Pecten robinaldinus D'Orb.
Pholadidea Dallasi Walk.
Thetis minor Sow.
Trigonia alæformis Park.

PLANTÆ[1].

Cladophlebis Albertsi Dunker
Pinites cylindroides Gardner
    „   *pottoniensis* Gardner
Fragments of coniferous and cycadaceous wood

LIST OF DERIVATIVE FOSSILS IN THE LOWER GREENSAND
OF POTTON.

(1) Miscellaneous derived fossils of doubtful horizon.

REPTILIA.

Dacosaurus sp.
Plesiosaurus sp.
Pliosaurus sp.

_____
[1] J. S. Gardner, *Geol. Mag.* Dec. 3, vol. III. (1886), p. 499.

PISCES.

> *Acrodus strophoides* McCoy
> *Edaphodon* sp.
> *Gyrodus* sp.
> *Hybodus* sp.
> *Pycnodus* sp.
> *Sphenonchus* sp.

MOLLUSCA.

> *Belemnites* sp.
> Various gasteropods
> *Cyprina* sp.
> *Lima* sp.
> *Modiola* 2 or 3 sp.
> *Myacites* sp.
> *Pholadomya* sp.

(2) From the Lower Greensand.

MOLLUSCA.

> *Ammonites Deshayesi* Leym.
> *Ancyloceras gigas* Sow. ?
> *Littorina* (several sp.)
> *Thetis minor* Sow.
> *Trigonia spinosa* Park. ?

(3) From the Wealden and Purbeck beds[1].

REPTILIA.

> *Iguanodon Mantelli* Meyer (bones, teeth, and scales of this
> genus very numerous)
> ,,   *bernissartensis* Boul.

PLANTÆ.

> *Endogenites erosa* Mant.
> *Kaidacarpum minus* Carr. (? derived)

(4) From the Portland beds.

MOLLUSCA.

> *Buccinum naticoideum* Sow.
> *Neritoma sinuosa* Sow.

[1] For particulars of reptilian remains see Prof. Seeley's *Index to Reptilia, etc. in Woodw. Mus.* (1869), pp. 74—80.

MOLLUSCA (*cont.*).

>*Arca* 2 sp.
>*Cardium dissimile* Sow.
>*Cytherea rugosa* Sow. ?
>*Lucina portlandica* Sow.
>*Sowerbya* sp.
>*Trigonia gibbosa* Sow.
>   ,,   *incurva* Sow.

(5) From the Kimeridge Clay.

REPTILIA.

>*Cimoliosaurus brachistospondylus* Hulke
>   ,,   *truncatus* Owen
>*Dacosaurus (Geosaurus) maximus* Plien.

PISCES.

>*Lepidotus maximus* Wagn.
>*Asteracanthus ornatissimus* Ag.

MOLLUSCA.

>*Ammonites biplex* Sow. (the commonest rolled fossil)
>   ,,   *cordatus* Sow. var. *serratus.*
>   ,,   *mutabilis* Sow. ?
>*Cardium striatulum* Sow.
>*Chemnitzia* sp.
>*Nucula ornata* Quenst. ?
>*Pleurotomaria* sp.

(6) From the Corallian.

ACTINOZOA.

>*Montlivaltia* ? sp.

BRACHIOPODA.

>*Rhynchonella varians* Schloth. ?

(7) From the Oxford Clay.

REPTILIA.

>*Cimoliosaurus trochanterius* Owen

MOLLUSCA.

Ammonites Lamberti Sow.

Gryphœa dilatata Sow. ?

(8) From some Oolitic rocks.

REPTILIA.

Megalosaurus Bucklandi Meyer

PISCES.

Strophodus magnus Ag.

ANALYSIS (I.) OF WASHED 'COPROLITES' FROM POTTON[1]
AND (II.) OF SIFTINGS.

						I.	II.
Water of combination		...	...			5·67	5·17
*Phosphoric acid		...	...		...	15·12	22·39
Lime	...	...	...	...	...	26·69	32·73
Mg. Al. and Fl.		...	...		...	4·51	6·64
†Carbonic acid	...		...	...	...	2·18	3·06
Iron oxide		...	...	...	...	20·61	8·08
Siliceous matter		...	...		...	25·22	21·93
						100·00	100·00
* Equal to tricalcic phosphate					...	32·76	48·51
† „ „ calcic carbonate					...	4·95	6·95

Wicken.

No workings have been open at this locality for many years. Mr Walker[2] mentions three beds of 'coprolites'; and Mr Keeping[3] gives another section in which two bands containing phosphatic nodules occur in the Lower Greensand and one in the overlying Gault. The Greensand rests on Kimeridge Clay and is overlaid

[1] Völcker, *Geol. Mag.* vol. III. (1866), p. 153.
[2] *Geol. Mag.* vol. IV. (1867), p. 309.
[3] *Geol. Mag.* vol. V. (1868), p. 272.

by Gault. The lower nodule bed is richest in fossils, indigenous and derived, and is often cemented into a hard conglomerate. In the sandy matrix in addition to the nodules there occur pebbles very similar in character to those at Potton. The nodules are of two colours, light and dark; the light ones are like those of Potton, while the others contain a smaller percentage of phosphate. Rolled fossils from the Lower Greensand of other areas, e.g. *Ammonites Deshayesi*, occur here as at Potton.

Upware[1].

The succession of the beds, which were here at one time exposed to view, is as follows (Fig. 4, p. 30). Lowermost is the Coral Rag of Upware; upon this rests Kimeridge Clay. Since the upper part of the Corallian knoll was stripped of its covering of Kimeridge Clay during the formation of the Lower Greensand the latter overlaps the denuded edges of the Clay and comes to rest on the limestone. The process of erosion which resulted in this relation of the beds caused the formation of a peculiar layer at the base of the Lower Greensand where it rests on the Coral Rag. This peculiar basal layer is composed of fragments of the Coral Rag imbedded in a matrix of worked-up Kimeridge Clay. Above this layer is a pebbly sand and rocky conglomerate about 1 ft. thick with nodules of phosphate of lime and many well-preserved mollusca. The pebbles are similar to those at Potton, and are mostly subangular in shape. Here and there the materials of the conglomerate are cemented together by carbonate of lime.

---

[1] W. Keeping, *The Fossils of Upware, etc.* (Sedgwick Essay, 1883); Bonney, *Camb. Geol.* Appendix i.

This lower nodule bed is succeeded by red and yellow sand. On this lies the upper nodule bed with paler-coloured nodules and little carbonate of lime. The pebbles are the same as in the lower nodule bed. Above this there is a layer of yellow sand, and then a bed of clay about 1 ft. thick which has been referred by some observers to the Gault. Resting on this is undoubted Gault with a nodule bed 5 inches thick at its base and rich in characteristic fossils. The total thickness of the Lower Greensand here is about 12 feet. Even in the immediate vicinity of Upware the nodule beds are impersistent and there may be only one present or as many as three.

The nodules themselves vary in colour from a dark chocolate to a pale yellow. They are sometimes recognisable casts of fossils, and sometimes irregularly shaped masses, and are often honeycombed by boring organisms.

In addition to these ordinary derived phosphatic nodules there are a few contemporaneously formed nodules consisting of a fine agglomeration of sand grains and phosphatic fragments, but these are accidental and of no commercial or scientific importance[1].

The Fossils. The fossils which are found in this series of sands and nodule beds of the Lower Greensand at Upware are partly derived and partly indigenous, as at Potton.

The indigenous fauna is large, consisting of more than 150 species (see list, p. 65), according to Keeping. The rich development of the Brachiopoda and of the large sponges and Polyzoa is a marked feature of the fauna. Some of the vertebrate remains are derived, others are

---

[1] W. Keeping, *The Fossils of Upware, etc.* p. 11.

probably indigenous, but the separation of these is very difficult[1].

The derived fossils which constitute the great majority of the nodules mostly belong to Jurassic species, and are much rolled and water-worn and pierced by boring organisms. They belong to species ranging from the Oxford Clay up to and including the lower beds of the Lower Greensand. Most of them are in the form of casts and are preserved in phosphate of lime, but the Oxford Clay fossils are in limonite and the wood is silicified, while the Corallian specimens have not suffered any chemical change. The vertebrate remains are more or less impregnated with iron or phosphate of lime. It is a remarkable fact that the fossils derived from early Lower Greensand beds are as thoroughly phosphatised and as much worn as those from the Jurassic beds. The phosphatising process must have gone on very quickly after the deposition of the sediments, for the horizon of some of these derived fossils is as high as that of the continental Aptian.

But in addition to the phosphatised Lower Greensand fossils there occur others as casts in ragged, eroded lumps of dark grit. Similar fossils in the same kind of rock occur at Potton and Hunstanton. It was held by Prof. W. Keeping[2] that these blocks are derived from the lower sands of the Lower Greensand similar to those at Donnington, in Lincolnshire, and that they are the sole remnants of a bed which, previous to the formation of the Upware, Potton and Hunstanton Lower Greensand, covered an extensive area.

[1] W. Keeping, *The Fossils of Upware, etc.* p. 30.
[2] *op. cit.* p. 37.

LIST OF NON-DERIVATIVE FOSSILS OF THE LOWER GREENSAND
OF UPWARE.

(From Prof. W. Keeping's list in *The Fossils, etc. of Upware*, 1883.)

REPTILIA.

Crocodilian teeth and
fragments of bones.
Dacosaurus sp.
Ichthyosaurus sp.

Iguanodon sp.
Plesiosaurus sp.
? *Pliosaurus* sp.

PISCES.

Acrodus sp.
Gyrodus sp.
Hybodus sp.
Ischyodus Townsendi Buckl.
Lepidotus maximus Wagn.
Otodus or *Oxyrhina* sp.
Pycnodus Couloni Ag.
Sphenonchus sp.
Strophodus (with *Asteracanthus*)

MOLLUSCA.

Cephalopoda.

Ammonites cornuelianus D'Orb.
,, *Deshayesi* Leym.
,, sp.
Ancyloceras Hillsi Sow.
Belemnites pistilliformis Blainv.
,, *subfusiformis* D'Orb.
,, *upwarensis* Keeping

Gasteropoda.

Cerithium marollinum D'Orb.
,, *neocomiense* Forbes
Littorina cantabrigiensis Keeping
,, *upwarensis* Keeping
,, *varicosa* Keeping
,, sp.
Nerinea tumida Keeping
,, sp.

MOLLUSCA (*cont.*).

 Gasteropoda (*cont.*).

  Patella sp.

  Pleurotomaria gigantea Sow.

     „ *Renevieri* Keeping

  Scalaria Keepingi Gardner

  Tessarolax Gardneri Keeping

  Tridactylus Walkeri Gardner

  Trochus n. sp.

  Turbo Reedi Keeping

Lamellibranchiata.

  Arca Carteroni D'Orb.

   „ *marullensis* D'Orb.

  Astarte subdentata Roem.

   „ sp. and n. sp.

  Avicula cornueliana D'Orb.

  Cardita rotundata P. and R.

  Cardium cottaldinum D'Orb.

   „ *subhillanum* Leym.

  Cucullœa subnana P. and R.

   „ sp.

  Cypricardia arcadiformis Keeping

   „ *squamosa* Keeping

   „ *striata* Gein.

  Cyprina angulata Flem. var. *rostrata* Sow.

   „ *obtusa* Keeping

   „ *Sedgwicki* Walker

  Exogyra Couloni D'Orb.

  Gryphœa dilatata Sow. ? (? derived)

  Lima faringdonensis Sharpe

   „ *longa* Roem.

  Lithodomus sp.

  Modiola obesa Keeping

   „ *pedernalis* Roem. ?

   „ sp.

  Neithea atava Roem.

   „ *Morrisi* P. and R.

   „ *ornithopus* Keeping

Mollusca (*cont.*).

Lamellibranchiata (*cont.*).

Nucula subtrigona K. and D.

Opis neocomiensis D'Orb.

Ostrea frons var. *macroptera* Sow.

,, *Walkeri* Keeping

Panopœa gurgitis D'Orb.

,, *plicata* Sow.

Pecten Dutempli D'Orb.

,, *orbicularis* Sow. var. *magnus* Keeping

,, *raulinianus* D'Orb.

Pectunculus marollensis Leym.

,, *obliquus* Keeping

,, *sublœvis* Sow.

Pholas (Fistulana) constricta Phil.

Plicatula œquicostata Keeping

,, *Carteroni* D'Orb.

Trigonia upwarensis Lyc.

Venus vectensis Forbes

Brachiopoda.

Rhynchonella antidichotoma Buv.

,, *cantabrigiensis* Dav.

,, *depressa* Sow.

,, *latissima* Sow.

,, *upwarensis* Dav.

Terebratella Davidsoni Walker

,, *Fittoni* Meyer

,, *Menardi* Lam.

Terebratula capillata D'Arch.

,, *Dallasi* Walker

,, *depressa* Lam.

,, ,, var. *cantabrigiensis* Walker

,, ,, ,, *cyrta* Walker

,, ,, ,, *uniplicata* Walker

,, *extensa* Meyer

,, *Lankesteri* Walker

,, *Meyeri* Walker

,, *microtrema* Walker

BRACHIOPODA (cont.).

Terebratula moutoniana D'Arch.
„   prœlonga Sow.
„   Seeleyi Walker
„   sella var. upwarensis Walker
Waldheimia celtica Mor. ?
„   Juddi Walker
„   pseudojurensis Leym.
„   tamarindus Sow. var. magna Walker
„   Wanklyni Walker
„   Woodwardi Walker

POLYZOA.

Ceriopora (Echinocava) Rawlini Mich.
„   (Reptomulticava) mamilla Reuss
Entalophora angusta D'Orb. ?
„   dendroidea Keeping
„   ramosissima D'Orb.
Heteropora arbuscula Keeping
„   coalescens Reuss
„   major Keeping
„   Micheleni D'Orb. ?
„   ramosa Röm.
„   (Nodicresis) annulata Keeping
„   (Reptonodicresis) sp.
Melicertites upwarensis Keeping
Radiopora bulbosa D'Orb. var.
Reptomultisparsa haimeana De Loriol
Semimulticava (Radiopora) tuberculata D'Orb.

ANNELIDA.

Serpula (Vermilia) ampullacea Sow.
„   antiquata Sow.
„   articulata Sow.
„   gordialis Goldf.
„   (Vermilia) lophioda Goldf.
„   plexus Sow.
„   rustica Sow.
Vermicularia Phillipsi Roem.
„   polygonalis Sow.

ECHINODERMATA.

> *Cidaris* sp.
> *Peltastes Wrighti* Desor.
> *Pseudodiadema rotulare* Ag.

PORIFERA.

> *Elasmostoma subpeziza* D'Orb. (*peziza* Goldf.)
> *Pachytilodia* sp.
> *Raphidonema* (*Catagma*) *cupuliformis* From.
>   ,,   *macropora* Sharpe (=*Elasmostoma acuti-*
>   *margo* Keeping, *non* Roem.)
>   ,,   *porcatum* Sharpe
> *Tremacystia* (*Verticillites*) *anastomans* Mant.
>   ,,   *annulata* Keeping
>   ,,   *clavata* Keeping

LIST OF DERIVATIVE FOSSILS IN THE LOWER GREENSAND
AT UPWARE.

(*a*) From the Portland beds.

MOLLUSCA.

> *Buccinum naticoides* Sow.
> *Astarte* sp. ?
> *Lucina portlandica* Sow.

(*b*) From the Kimeridge Clay.

> (These form the majority of the derived fossils.

REPTILIA.

> *Ichthyosaurus* sp.
> *Plesiosaurus* sp.
> *Pliosaurus brachydeirus* Owen

PISCES.

> *Acrodus* sp.
> *Asteracanthus* sp.
> *Gyrodus* sp.
> *Hybodus* sp.

PISCES (*cont.*).

Lepidotus sp.
Otodus sp.
Pycnodus sp.
Sphenonchus sp.
Strophodus sp.
Chimæroids

MOLLUSCA.

Ammonites biplex Sow.
" Kœnigi Sow.
Belemnites explanatus Phil.
Cardium striatulum Sow.
Myacites sp.
Pleurotomaria reticulata Sow.
Trochus sp.

ANNELIDA.

Serpula tricarinata Sow.

(c) From the Coral Rag.

(N.B. These are not phosphatised.)

MOLLUSCA.

Cerithium muricatum Sow.
Chemnitzia heddingtonensis Sow.
Exogyra sp.
Gryphœa dilatata Sow. var. ?
Lithodomus inclusus Phil.
Opis corallina Dam.
Ostrea gregaria Sow.
Pecten vimineus Sow.
Unicardium sp.

ECHINODERMATA.

Apiocrinus sp. ?
Cidaris florigemma Phil.
Echinobrissus scutatus Gmel.
Glypticus hieroglyphicus Goldf.
Hemicidaris intermedia Flem.
Holectypus depressus Lam.
Pseudodiadema hemisphericum Ag.

(*d*) From the Oxford Clay.

(N.B. These are not phosphatised but are preserved
in limonite.)

MOLLUSCA.

Ammonites biplex D'Orb. *non* Sow.
Arca sp.
Modiola sp.

(*e*) From Lower Greensand beds.

MOLLUSCA.

Ammonites Deshayesi Leym.
„ sp.
Ancyloceras sp.
Hamites sp.
Cerithium sp.
Littorina (several sp.)

BRACHIOPODA.

Terebratula ovoides Sow.
„ „ var. *rex* Lank.

(*f*) From a dark grit of Lower Greensand age.

MOLLUSCA.

Cerithium granulatum Phil. ?
Solarium neocomiense D'Orb.
Trochus sp.
Cucullæa vagans Keeping
„ *donningtonensis* Keeping
Mytilus lanceolatus Sow.
Pecten orbicularis Sow.
Perna ricordeana D'Orb.
„ *Mulleti* Sow.
Thetis minor Sow.
Trigonia vectiana Lyc.
„ sp.

BRACHIOPODA.

Terebratula ovoides Sow. ?
„ „ var. *rex* Lank. ?

Mode of formation of the beds. The action of strong currents in shallow water and the rapid accumulation of the deposits are shown by the presence of coarse conglomerates, subangular pebbles, and false-bedding. The character of many of the pebbles (quartzites, lydianites, etc.) points to their derivation from Palæozoic rocks. The variety of formations from which the water-worn fossils come indicates that the land from which they were washed down was composed of almost every member of the Upper Jurassic, while the presence of fossils derived from early Lower Greensand beds makes it evident that there had been a recent alteration in the level of the sea-floor, bringing newly-formed beds within the range of the denuding and transporting action of waves and ocean currents.

The geographical position of these abnormal Lower Greensand deposits which we meet with at Potton, Upware and Hunstanton, taken together with the foregoing considerations, leads us to the view that they accumulated in a narrow channel which connected the Speeton and North-German and Anglo-Gallic basins. This channel was formed in late Lower Greensand times by earth movements which shallowed the water over a considerable area and converted a portion of the open sea in the Eastern Counties into narrow straits swept by powerful currents and flanked on the east by a ridge chiefly of Palæozoic rocks, which is now only reached in deep borings, and on the west by banks of soft Jurassic clays and limestones.

Correlation of the beds with those of other areas. A close resemblance exists between the sands and pebble beds of Potton, Upware and Hunstanton, and the Sandgate and Hythe Beds of Kent and Surrey. The Atherfield Clay has a totally different fauna and was formed much earlier, but still in Lower Greensand times.

Certain parts of the sands, limestones and clays of the Lower Greensand of the country around Godalming[1] show a close resemblance both in fauna and in lithological characters to those of our area. Thus the ten determinable species of fossils in the " Middle Series " of Meyer all occur in the Upware deposits and at Brickhill; and the pebble bed at the base of Meyer's " Upper Series " has a fauna of which every species occurs in the Cambridgeshire bed. In addition to this palæontological identity the rock-characters are very similar, for the Godalming bed consists of a coarse sand with angular pebbles of quartzite, lydianite, jasper, etc., as well as similar phosphatic nodules and derived Jurassic fossils. The large majority of the fossils in the overlying "Bargate Stone" also occur in the Upware beds. The Shanklin Sands of the Isle of Wight and the Hythe Beds of East Kent, particularly the upper part, show a close relationship to our representatives of the Lower Greensand.

Mode of formation of the phosphatic nodules. As above stated many of the nodules are easily recognised to be of organic origin, and to consist of the phosphatised casts of the shells of mollusca, or the bones of reptiles and fish. In some cases the nodules are merely shapeless lumps, and their organic origin can only be suspected by analogy. It has been thought by many writers on the subject that the fossils must have been in the phosphatic condition before they were washed out of

[1] C. J. A. Meyer, *Proc. Geol. Assoc.* 1869. (Separate Paper.)

their parent formation.   But although indigenous phosphatic nodules indeed occur in the Jurassic and other
rocks from which these phosphatised organisms are derived, yet it is very rare to find any fossils in this condition in the parent rock.   It seems in fact necessary to
suppose that the replacement of the original matter by
phosphate of lime must have taken place in most cases
in the Lower Greensand sea by a kind of pseudomorphing
process.   The identical condition and character of the
derived phosphatised organisms, whether they come from
the Kimeridge Clay, the Portland series, the lower part
of the Lower Greensand, or other beds, seem to support
this view.   The two facts that the Oxford Clay fossils consist
of limonite, and that the Coral Rag ones are unaltered,
are exceptions difficult to explain, but their condition may
be due to some original want of susceptibility to the phosphatising chemical influences.   One condition which seems
necessary for such a phosphatising process to take place
is long exposure of the organic remains on a sea-bottom
where no sediment is being deposited.

THE UPPER CRETACEOUS.

THE GAULT.

General remarks. The Gault of Cambridgeshire, and parts of Bedfordshire and Hertfordshire, instead of showing evidence of a gradual passage upwards into the overlying beds, as in other areas, seems to have suffered denudation of its upper half prior to the deposition of the Chalk Marl. The Cambridge Greensand contains the result of the washing, sifting and rearranging of the materials of the Gault which suffered this erosion.

In addition to this feature of contemporaneous erosion, the Gault of this and the neighbouring areas is of special interest on account of its change in lithological characters and in thickness when traced along its strike. The question of its relation to the Red Chalk of Hunstanton must also be considered in this connection, especially since Hunstanton is within easy reach of Cambridge.

Characters and thickness[1]. The Gault is divisible into two main zones—an upper and a lower—of which the former is sometimes called the zone of *Ammonites inflatus*, and the latter the zone of *Am. lautus*. The Upper Gault is absent in Cambridgeshire, having been eroded away before the deposition of the Chalk Marl; and probably part of the Lower Gault is also wanting.

[1] *Mem. Geol. Surv. Explan. Quart. Sheet* 51, S.W. p. 13 *et seq.* A. J. Jukes-Browne, *Q. J. G. S.* vol. xxxi. (1875), p. 256 *et seq.* A. J. Jukes Browne and W. Hill, *Q. J. G. S.* vol. xliii. (1887), p. 593 *et seq.*

There is also a natural thinning of both divisions of the Gault to the north-east. Thus at Totternhoe in Bedfordshire and at Stoke Ferry in Norfolk both divisions of the Gault are present in their entirety, but while at Totternhoe the Gault is 230 feet in thickness, at Stoke Ferry it is only 60 feet, this difference of 170 feet being the result of thinning-out.

Between Tring and Hitchin the Lower Gault increases in thickness from 150 feet to 204 feet, but the Upper Gault decreases in thickness from about 80 feet to 20 or 30 feet. North of Arlesey the Lower Gault begins to grow thinner, and it is assumed that this is mainly due to the erosion it has undergone; but natural thinning-out also of the division in a northerly as well as a southerly direction appears to commence about this point. Between Arlesey and Ashwell the total thickness of the Gault ranges from 180 to 200 feet. On the assumption that at or near Arlesey the Lower Gault had its greatest thickness, and that the original surface-slope of the bed before erosion was regular and constant to the north as far as Stoke Ferry, this division should have a thickness of about 150 feet at Cambridge. To this must be added a certain thickness for the Upper Gault—about 20 feet, according to Mr Jukes-Browne. This would give 170 feet as the original thickness of the whole Gault at Cambridge. But since well-borings give on an average only about 125 feet of Gault at Cambridge, we must conclude that some 50 feet have been washed away.

There is evidence that the base of the Gault is irregular, and that it fills up local troughs and hollows on the surface of the Lower Greensand. The supposed unconformable transgression of the Gault over the Lower Greensand has already been mentioned (p. 49); and the statement

that some borings in Cambridge have pierced through
from 130 to 200 feet of Gault would, if confirmed, make it
necessary to imagine that the base is here very uneven.

Around Reach and Burwell the Gault is about 100
feet thick: at Soham it is 90 feet thick.

If we look at the present slope of the surface of the
land and have in our mind's eye the slope of the original
surface of the top of the Gault prior to its erosion in
Cambridge Greensand times, we shall see that it is
probable that some 50 feet or so of Gault have been
removed from over this district. It is over this area of
erosion that the Cambridge Greensand extends. Where,
on the other hand, the Gault has suffered no erosion but
is present in its completeness, there the Cambridge
Greensand is absent.

According to the diagram given by Jukes-Browne[1] the
prolongation northwards of the present slope of the land
beyond Soham meets the line from the base of the Upper
Gault about seven miles south of Stoke Ferry; and hence
it may reasonably be supposed that the Cambridge Greensand extends to this point, but it has not been actually
seen more than 2 miles north-east of Soham.

At Stoke Ferry the Gault is 58 feet in thickness, and
as far north as this locality the formation preserves the
character of a tough blue clay. Above it at this spot
there lies a glauconitic marl, into which it is probable that
the Cambridge Greensand of the south passes horizontally.
A dark green sand forms the base of the Gault, and at
West Dereham this sand contains numerous phosphatic
nodules and rests on an unevenly eroded surface of brown
sand, that is, the top of the Lower Greensand. At Roydon,
a little further north, in the neighbourhood of Grimston,

[1] *Q. J. G. S.* vol. XLIII. (1887), p. 595.

the Gault has diminished to about 20 feet in thickness and consists of a lower division of blue clay, a thin middle layer of yellowish marl with reddish blotches, and an upper division of tough grey marly clay. Thin yellowish limestone bands are found in it also near Grimston. At Dersingham, still further north, below the Chalk Marl come 2 feet of softish pale-grey marl, and beneath this are 3 feet of yellowish-brown chalky material which passes down into 2½ feet of red marly clay. These thin beds of marly clay and limestone are the attenuated representatives of the thick blue clay which constitutes the Gault further south.

The Norfolk Gault is not only more calcareous than that of the Midlands, but also contains as it is followed northwards a continually decreasing proportion of mechanical sediment, such as quartz-grains; on the other hand the organic material (Foraminifera, shell-fragments, etc.) steadily increases. The malmstones and sands of the so-called Upper Greensand of the southern counties are entirely wanting, and everything points to the conclusion that we are getting further away from the old coast-line and from the source of the mechanical sediment, and nearer to deep-water and pelagic conditions.

The phosphatic nodules occur frequently in layers at various horizons, or scattered irregularly through the Gault. They are often associated with broken and rounded fragments of fossils; but they do not necessarily point to a great break in the sequence, but only to an interruption in sedimentation or to a sifting and sorting by current action of a certain thickness of clay which contained semi-phosphatised and semi-consolidated casts of fossils. Ammonites in the state of clay casts, with only their inner whorls preserved in solid phosphate, have been found in the Gault at Muzzle by Messrs Jukes-Browne and Hill[1], and a gentle washing of these casts would remove all the clayey portion and leave only the solid phosphatic portion behind. We may thus account for the numerous fragmentary and imperfect fossils in the nodule layers of the Gault. But as in the case of the Lower Greensand a cessation of the deposition of sediment on the sea-floor appears to be the main condition necessary for the activity of the phosphatising process.

These nodule beds occur at various horizons, *e.g.* near the base of the Gault at Wendy and Wimpole, and at West Dereham in Norfolk[2], and at 60 or 70 feet down in the Gault at Whaddon, Guilden Morden, Ashwell and Harlton.

The phosphatic nodules themselves are of a dark colour but are paler than those from the Cambridge Greensand.

The so-called 'rugg-stones' are merely decomposing lumps of iron pyrites.

Occurrence. The lower boundary-line of this formation is quite hidden by Boulder Clay for a long distance in West Cambridgeshire, but its upper limit is well known by means of the 'coprolite' pits which have been worked all along its junction with the Cambridge Greensand. Owing to the three promontories of Chalk, which stretch out eastwards from the great western outlier of Upper Cretaceous beds, and are only separated from the main mass in the east by the valleys of the Rhee and the Cam, the Gault plain is divided up into four fairly distinct areas. Moreover, as a result of the presence of these projecting masses of Chalk and the patches of drift and

---

[1] *Q. J. G. S.* xliii. (1887), p. 572.

[2] *Mem. Geol. Surv. Explan. Sheet* 65, p. 22.

river-gravel, the continuity of the valley which extends
through the southern midland counties between the top of
the Lower Greensand and the base of the Chalk is not so
apparent in Cambridgeshire. Of the four chief areas of
Gault in the county the most southern lies to the south of
the Orwell-Barrington ridge. The second area extends
from the north side of this ridge to the Barton ridge of
Chalk ; and still further north are the two remaining areas,
of which the more southern and smaller one is scarcely
defined on the north by the discontinuous Madingley-
Coton and Observatory ridge. It is on this smaller area
that the town of Cambridge mainly stands. A small part
of this area lies on the east bank of the river Cam, but
much of the outcrop near the river is obscured by the
Pleistocene gravels[1].

Finally, the large northern area extends up to the
outcrop of the Lower Greensand in the north, and its
eastern boundary is marked by the course of the river Cam
nearly as far as Waterbeach, beyond which the main outcrop lies to the east of the river. Though much hidden
by gravels between Chesterton and Denny Abbey yet it
reappears as a long strip between Waterbeach and Causeway End Farm on the west side of the Cam. It underlies
the fen south of Upware, the small patch to the west of
the Corallian mass being probably an outlier. Near
Wicken and Soham the Gault again reaches the surface,
and surrounds the plain of Soham Mere as a horseshoe-
shaped ridge. To the north-east of this it again sinks
below the fens, but emerges near Stoke Ferry and thence
runs northwards, more or less continuously at the surface,

---

[1] The inliers of Gault near Haslingfield, Hauxton Mill Bridge and
High Hall Farm at Horningsea, are due to the irregularity of its upper
surface, which at these places forms ridges or hillocks.

with the remarkable change in thickness and characters already described, to appear at Hunstanton as the Red Chalk.

Economics. The Gault at Cambridge is extensively dug for bricks, tiles, pipes, etc. Some of the largest pits are at Barnwell; another is at the beginning of the Barton road. Near Chesterton, Impington, and Clayhithe are other brickyards. The bricks made from the Gault are yellow in colour, except where the iron in the clay has been oxidised by currents of air admitted through cracks in the clamp—in which case the bricks are red. Phosphatic nodules—'coprolites'—occur in the clay at many places (Barnwell, Chesterton, etc.).

The Gault forms a stiff cold soil.

Exposures and special sections. In the Barnwell and other brick-pits good sections of the Gault are to be seen. Its upper portion is exposed in most of the 'coprolite' diggings in the Cambridge Greensand and its lower boundary is seen in a pond near the farm S.E. of Gamlingay Station.

Palæontology. It must be remembered that the fossils are all Lower Gault forms, the Upper Gault having been eroded away from this area, as explained above. The fossils are scarce in this formation near Cambridge, but they are said to be more plentiful near its base than in the higher parts. In the Woodwardian Museum there are the following species from Barnwell, Ely, and Upware :—

REPTILIA.

Ichthyosaurus sp.
Rhinochelys sp.

R.   6

PISCES.

    Cimolichthys (Saurocephalus) striatus Ag.
    Ptychodus sp. (dorsal spine)

MOLLUSCA.

  Cephalopoda.

    Ammonites interruptus Sow.
        „   *raulinianus* D'Orb.
        „   *rostratus* Sow.
        „   *serratus* Park.
        „   *varicosus* Sow.
    Belemnites attenuatus Sow.
        „   *minimus* List.
    Hamites sp.

  Gasteropoda.

    Cerithium ornatissimum Sow.
    Dentalium ellipticum Sow.

  Lamellibranchiata.

    Inoceramus concentricus Park.
        „   *sulcatus* Park.
    Neithea quinquecostata Sow.
    Nucula ovata Mant.
        „   *pectinata* Sow.
    Ostrea sp.
    Perna sp.
    Plicatula pectinoides Sow.

BRACHIOPODA.

    Terebratula biplicata Broc.

ECHINODERMATA.

    Pentacrinus Fittoni Aust.
    Pseudodiadema sp. (spines)

ACTINOZOA.

    Trochocyathus angulatus Dunc.

THE RED CHALK OF HUNSTANTON [1].

Since Hunstanton is so frequently visited by Cambridge students, a description of the famous Red Rock and its relations to other beds will be useful.

The cliffs at Hunstanton show three bands of colour; the uppermost one is white and is the broadest of the three; the middle one is red; and the lowest one is yellowish-brown. These three colour-bands are formed respectively by the lower part of the Lower Chalk, the Red Chalk, and the Carstone or Lower Greensand.

The Carstone consists of ferruginous sandstones and fine conglomerates. A little west of the lighthouse more than half the height of the cliffs is composed of these arenaceous beds. The middle member of the Carstone here seen is a compactly cemented, fine ferruginous conglomerate of small rounded and subangular pebbles, and is about four feet in thickness. When the easterly dip of the beds brings this conglomerate down to the level of the beach near the lighthouse, it forms a platform of rounded or subquadrangular blocks, due to the weathering along its joint lines (see Fig. 7). Blocks half quarried out of the base of the cliff by this weathering process may be seen here. All this series assumes a much darker colour when acted upon by the sea-water.

Above the softer yellow sandstone, which overlies this conglomerate to a thickness of about five feet, comes the Red Chalk. This bed is three feet thick and is divisible into three more or less distinct layers [2]:—

---

[1] For detailed bibliography see W. Whitaker, Presid. Address, *Proc. Norwich Geol. Soc.* part VII. 1883.

[2] Wiltshire, *Q. J. G. S.* vol. xxv. (1869), p. 185.

    1. Hard lumpy reddish chalk or mottled red and white.

    2. Rough nodular red limestone passing down into

    3. Deep-red gritty rock, softer at the base.

FIG. 7. BASE OF THE CLIFF AT HUNSTANTON.

a. Hard creamy white chalk (CHALK MARL) passing gradually down into

b. Gritty grey hard chalk (CHALK MARL) containing fragments of *Inoceramus* and green-coated nodules. The latter form a layer at its base. Thickness 4 feet.

c. Hard white chalk (the "Sponge-bed"). Thickness 1½ feet.

d. RED CHALK. Thickness 3 feet, divisible into three portions :—

    1. Hard lumpy reddish or mottled red and white chalk.

    2. Rough nodular red limestone passing down into

    3. Deep-red gritty rock, softer at the base.

e. Ferruginous brown sandstone and fine conglomerates. The rounded blocks *e'* represent a harder bed weathered out *in situ*. Some masses of white chalk, fallen from the cliff above, rest upon its surface.

Immediately overlying the Red Rock is a white nodular limestone—the so-called Sponge-bed—about one foot and a half thick.

Above this comes the rest of the Chalk Marl consisting in its lower part of about four feet of hard grey and very gritty chalk containing green-coated nodules and very abundant fragments of *Inoceramus*. The upper part is of hard creamy-white chalk, 13 feet or so in thickness.

Near the lighthouse a band of hard gritty dark grey chalk, two feet thick, is seen overlying the Chalk Marl. This represents the Totternhoe Stone with which we are familiar at Burwell and elsewhere. Above this band come nine feet of the Zone of *Holaster subglobosus*, which is thin-bedded and has some indefinite marly bands near its base. The cliffs are capped with soil and chalk rubble[1] (Fig. 7).

Composition of the Red Chalk. From the chemical analysis of samples of the Red Chalk we find that it is a marly limestone with a high percentage of peroxide of iron which forms the colouring matter. It contains scarcely any fine mechanical sediment, but small pebbles of quartz, slate, etc. are scattered throughout its mass. Occasionally these pebbles are of large size.

The source of the peroxide of iron which colours the bed has been ascribed by some writers to the decomposition and peroxidisation of green grains of silicate of iron in the bed, such as are now present in the Gault of the southern counties. It has been pointed out that the amount of iron in the Hunstanton rock is really less than

[1] For further details and references see Wiltshire, *Q. J. G. S.* vol. xxv. (1869), p. 185; Bonney, *Camb. Geol.* (1872), App. III.; *Records Geol. Assoc. Excursions* (1891), p. 202.

in the Gault of Folkestone, but that in the former the iron is present in the state of sesqui-oxide (red oxide) while in the latter it is all in the state of protoxide. If the Gault of Folkestone were oxidised it would assume the colour of the Hunstanton rock.

It has been suggested that the iron was introduced from the Carstone below by capillary attraction into the Hunstanton rock subsequent to its formation, but there does not seem to be sufficient ground for this view. Nor is there any substantial evidence in favour of regarding the rock as corresponding to the 'red clay' which is now forming in the abysses of the Atlantic and Pacific Oceans.

Analysis of Red Chalk, quoted by Prof. Bonney[1].

Silica	9·28
Carbonate of lime	80·04
Sulphate	0·10
Peroxide of iron	9·60
Alumina	1·42
Manganese	trace
	100·44

Analysis of pink sample of Red Chalk from Hunstanton, by Dr W. Johnstone[2].

Silica, etc.	7·50
Carbonate of lime	83·81
Peroxide of iron	5·72
Alumina	1·67
Manganese	0·58
Magnesia	0·62
	99·90

[1] *Geologist* (1863), p. 29.
[2] *Q. J. G. S.* vol. XLIII. (1887), p. 588.

Analysis of the Red Marl from Grimston (A) *and from Dersingham* (B) *by Dr W. Johnstone*[1].

	(A)	(B)
Silica and silicates	... 22·60	25·70
Carbonate of lime	... 69·50	64·49
„ „ magnesia ...	... 0·90	1·32
Sulphate of lime	... 0·66	0·33
Peroxide of iron	... 3·40	4·16
Alumina and phosphoric acid	... 1·60	0·80
Manganese	... trace	trace
Organic matter, etc.	... 1·34	3·20
	100·00	100·00

Stratigraphical relations and age. The absence at Hunstanton of any bed like ordinary Gault, or Upper Greensand, and of any rock exactly resembling the Chalk Marl has led different observers to refer the Red Chalk to one of these three formations. On the strength of recent work[2] the most satisfactory view appears to be that the Red Chalk is the diminutive representative of the whole of the Gault of other areas. The normal state of things north-east of Buckinghamshire is a sudden transition from Upper Gault to Chalk Marl without the intervention of sandy beds such as can with any certainty be classed with the Upper Greensand of the south. This is owing to the change in the conditions of deposit as we proceed to the north-east, for we get away from the zone of terrigenous sediments and enter the area where oceanic deposits accumulated.

The gradual transition from littoral to pelagic conditions is interrupted over the tract where the Cambridge

[1] *Q. J. G. S.* vol. XLIII. (1887), p. 588.
[2] Jukes-Browne and Hill, *Q. J. G. S.* vol. XLIII. (1887), p. 592.

Greensand is found, but this deposit is due to an excep-
tional contemporaneous denudation of lower beds (see p. 95
et seq.) and does not upset the general conclusion drawn
from the horizontal change in the characters of the Gault
and overlying beds. At Hunstanton therefore we do not
need to look for typical Upper Greensand deposits possessing the littoral characters of those in the south of England ;
for these shore-deposits have thinned out long before
Hunstanton is reached.

The age of the Red Chalk is therefore considered to
be either that of the Gault or of the Chalk Marl, and if
we bear in mind the gradual change in the characters and
thickness of the Gault when traced northwards through
Norfolk, we shall be inclined from these reasons alone to
correlate the Red Chalk with the Gault. As already
described (p. 78), the latter grows more and more calcareous
and freer from inorganic material and at the same time
decreases in thickness in a northern direction. Thus at
Dersingham it is represented by seven feet of marly and
chalky material, the lower part of which is coloured red.
The Chalk Marl also alters in character by becoming harder
and less marly, and at the same time steadily grows
thinner. There is a hard whitish limestone directly overlying the Gault at Roydon and other places inland which
is identical with the " Sponge-bed " immediately resting
on the Red Chalk in the Hunstanton cliffs.

Microscopically, also, the structure of the Red Chalk
bears the same relation to the red and yellow marls of
Dersingham that the hard Chalk Marl of Norfolk does to
the softer Chalk Marl of Cambridgeshire.

The palæontological evidence supports the conclusions
derived from the consideration of the physical characters
and relations of the deposit.

Fauna of the Red Chalk. The fauna has somewhat mixed characters and affinities.

Firstly, almost all the fossils of the Gault of West
Norfolk have been found in it, and amongst them are
eight characteristic Gault species of Ammonites (*Am.
auritus, Am. Beudanti, Am. interruptus, Am. lautus, Am.
ochetonotus, Am. rostratus, Am. splendens,* and *Am. tuber-
culatus*). On the other hand the Ammonites characteristic
of the Chloritic and Chalk Marl are conspicuously absent.

Secondly, there are some fossils other than Ammonites
which in other regions belong to higher beds than the
Gault. These are chiefly pelagic species of the White
Chalk, and as the Red Chalk shows itself in lithological
characters to be a deeper sea deposit than the common type
of Gault we should naturally expect the presence of such
forms. The great rarity of Gasteropods in the Red Chalk
though they are so abundant in the Gault of the south
of England finds a similar explanation, for Gasteropods
flourish best in shallow water.

Thirdly, the state of preservation of the fossils ab-
solutely excludes the idea that they can be derived, yet
some of the fossils belong to the Lower Gault of other
areas.

The deeper water, almost unpolluted by mechanical
sediment, in which the Red Chalk is shown by its lithological and faunistic characters to have accumulated, would
support many forms of life which could not survive in the
shallow and muddy waters in which the blue Gault clay
was deposited. An alteration in physical conditions and
general environment must always be expected to have
a more or less marked effect on the fauna, and a change
in the composition and character of a rock is therefore
always accompanied by some differences in the fossils

contained in it. Such an alteration appears to be here evident; and the mixed Upper and Lower Gault species which constitute the bulk of the fauna, the absence of many shallow-water forms, and the occurrence of some deeper water forms, together with the stratigraphical evidence of the position of the bed and the lithological indications of its mode of formation lead us to conclude that the Red Chalk of Hunstanton is the condensed representative of the whole of the Gault of other areas and was formed under pelagic conditions outside the reach of mud-bearing currents.

LIST OF FOSSILS FROM THE RED CHALK.

FORAMINIFERA. (See *Journ. Roy. Microsc. Soc.* part v. p. 549, 1890.)

PORIFERA.

> *Scyphia tenuis* Roem.
> *Spongia paradoxica* Webster (? organic[1])

ACTINOZOA.

> *Cyclolites polymorpha* Goldf.
> *Micrabacia coronula* Goldf.
> *Podoseris elongata* Duncan
>   ,,   *mamilliformis* Duncan

ECHINODERMATA.

> *Bourguetticrinus rugosus* Ag.
> *Cidaris gaultina* Forbes
>   ,,   *vesiculosa* Wright
> *Holaster suborbicularis* Ag.
> *Peltastes Wiltshirei* Seeley
> *Pentacrinus Fittoni* Aust.

---

[1] Prof. Hughes argues for its inorganic nature, *Q. J. G. S.* XL. (1884), p. 273.

ECHINODERMATA (*cont.*).

Pseudodiadema Brongniarti Ag.
      „   *ornatum* Goldf.
Torynocrinus canon Seeley

ANNELIDA.

Serpula antiquata Sow.
    „   *cristata* Duj.
    „   *depressa* Goldf.
    „   *rustica* Sow.
Vermicularia umbonata Mant.

CRUSTACEA.

Pollicipes unguis J. Sow.

BRACHIOPODA.

Kingena lima Defr.
Rhynchonella sulcata Park.
Terebratula biplicata Broc.
     „   *capillata* D'Arch.
     „   *semiglobosa* Sow.
Terebratulina gracilis Schl.

POLYZOA.

Berenicea regularis D'Orb.
Proboscina dilatata D'Orb.
Reptomulticava mamilla Reuss
Stomatopora longiscuta D'Orb.

MOLLUSCA.

Cephalopoda.

Ammonites auritus Sow.
     „   *Beudanti* Brong.
     „   *interruptus* D'Orb.
     „   *lautus* Sow.
     „   *ochetonotus* Seeley
     „   *rostratus* Sow.
     „   *splendens* Sow.
     „   *tuberculatus* Sow.
Belemnites attenuatus Sow.

MOLLUSCA (*cont.*).

    Cephalopoda (*cont.*).

      Belemnites minimus List.

            ,,   *ultimus* D'Orb.

      Nautilus albensis D'Orb.

    Gasteropoda.

      Cerithium mosense Buv.

      Pleurotomaria sp.

    Lamellibranchiata.

      Avicula gryphæoides J. Sow.

      Exogyra conica Sow.

        ,,   *haliotoidea* Sow.

        ,,   *laciniata* Nils.

        ,,   *rauliniana* D'Orb.

      Inoceramus Crispi Mant.

        ,,   *subsulcatus* Pictet and Roux

        ,,   *sulcatus* Sow.

        ,,   *tenuis* Mant.

      Lima globosa J. Sow.

        ,,   *iteriana* P. and R.

      Ostrea normaniana D'Orb.

        ,,   *vesicularis* Lank.

      Pecten Beaveri Sow.

      Plicatula pectinoides Sow.

      Spondylus striatus Sow.

PISCES.

    Ischiodon sp.

    Lamna appendiculata Ag.

REPTILIA.

    Plesiosaurus latispinus Owen?

THE CHALK.

General remarks. The well-known series of soft white earthy limestones, called the Chalk, is divided up and classified partly by lithological characters and partly by fossil zones.

The classification adopted for the beds in Cambridgeshire is as follows:—

Upper Chalk {Zone of *Micraster cor-testudinarium*.

Middle Chalk ⎨ Chalk Rock or Zone of *Heteroceras reussianum*.
Zone of *Holaster planus*.
Zone of *Terebratulina gracilis* or 'Wandlebury beds.'
Zone of *Rhynchonella Cuvieri*, with Melbourn Rock at base.

Lower Chalk ⎨ Zone of *Belemnitella plena*, or 'Belemnite Marls.'
Zone of *Holaster subglobosus* with Burwell Rock at base.
Zone of *Ammonites varians*, or the Chalk Marl, with the Cambridge Greensand at base.

There are three beds in this series which possess remarkable persistence in lithological character throughout the district. They are the Burwell Rock, the Melbourn Rock, and the Chalk Rock. The Burwell Rock, long known under that name in this district, was described by Mr Whitaker under the name of Totternhoe Stone in Beds. and Bucks.[1] The Melbourn Rock[2] marks the base

---

[1] Whitaker, *Q. J. G. S.* vol. xxi. (1865), p. 398; *Mem. Geol. Surv.* vol. iv.

[2] *Mem. Geol. Surv. Explan. Quart. Sheet* 51, S.W. p. 55; A. J. Jukes-Browne and W. Hill, *Q. J. G. S.* vol. xlii. (1886), p. 216.

of the Middle Chalk, and the Chalk Rock[1] the top of the same division.   The so-called Wandlebury beds[2] between the two rock-beds last mentioned are not so important or persistent as a lithological horizon.

THE LOWER CHALK.

General remarks.   The Lower Chalk comprises the beds commonly known as the Chalk Marl, the Grey Chalk and the Belemnite Marls.   Until the separation of the latter from the Melbourn Rock and their addition to the top of the Lower Chalk[3], all the upper portion of the Lower Chalk was called the Grey Chalk, but since it has been recognised that this term is lithologically inappropriate and incapable of general application the subdivision of the beds by means of palæontological zones has been usually adopted.   The nomenclature employed by Messrs Whitaker and Jukes-Browne in their "Geology of London" (*Mem. Geol. Surv.* 1889) is here followed.

The Chalk Marl in this area may practically be said to consist of two divisions, *i.e.* (1) the Cambridge Greensand at the base; and (2) the Chalk Marl proper.

These two divisions are not sharply marked off from each other, for the basal layer, which is generally under a foot in thickness, graduates up imperceptibly into the overlying chalky beds.   The importance of this basal layer—the Cambridge Greensand—is so great that it must be described separately.

[1] W. Whitaker, *Q. J. G. S.* vol. XVII. (1861), p. 166 ; *Mem. Geol. Surv.* vol. IV.

[2] *Mem. Geol. Surv. Explan. Quart. Sheet* 51, S.W. p. 63.

[3] W. Hill and A. J. Jukes-Browne, *Q. J. G. S.* vol. XLII. (1886), p. 216.

(1) THE ZONE OF *AMMONITES VARIANS*.

(a) THE CAMBRIDGE GREENSAND.

General remarks. This is geologically the most important and interesting bed in the whole area. Its fauna, lithological peculiarities, mode of occurrence and manner of formation, demand close attention.

Characters and thickness. The Cambridge Green-sand is a greenish clayey or sandy marl containing abundant glauconitic grains which give it its colour and its sandy texture. In places the large admixture of Gault mud imparts a more clayey character and darker tint. It contains in its lowermost layer abundant dark olive-black or brownish black nodules largely composed of phosphate of lime. These are the 'coprolites[1]'—erroneously so called—for which the bed is commercially valuable. It is crowded also with fossils which have the same composition and appearance as the nodules. These fossils are usually in a rolled and water-worn condition, for they are derived from earlier formed beds.

But there is also an indigenous fauna represented by more perfect and paler coloured fossils. These are however less numerous and do not occur in the very lowest layer. With the indigenous fauna are associated lighter coloured nodules which range up together with most of this fauna into the overlying Chalk Marl proper.

In addition to these nodules, boulders of far-travelled rocks are frequently found. They are rounded or angular

---

[1] From the Greek κόπρος, dung, a name given when they were supposed to be the fæcal remains of fish and reptiles. It is now known that true coprolites, though not uncommon in the Chalk, form but a small proportion of the phosphate nodules known in commerce as coprolites.

in shape, and vary from about one inch to more than a foot in diameter (see p. 103).

The actual nodule bed is usually about 6 inches in thickness and rarely exceeds 10 inches, but the Greensand or a glauconitic marl in which scattered nodules occur generally extends up for another five or six inches before merging insensibly into the genuine whitish Chalk Marl. When the surface of the underlying Gault is very un-even, the Greensand is always thinner on the top of the ridges and thicker in the intervening hollows, being frequently 18 to 24 inches thick in the latter; the basal nodule-bearing layer generally exhibits a similar thinning and thickening, but occasionally is absent altogether on the top of the ridges[1] (Fig. 8).

FIG. 8.   SECTION SEEN IN PHOSPHATE DIGGINGS CLOSE TO QUAKER'S MILL, BARRINGTON, *Oct.* 11, 1883.   Scale 10 ft. to 1 inch.

(By Prof. T. McKenny Hughes.)

a.   Surface soil, rainwash, &c.

b[1].   River silt with fragments of clunch, &c.
b[2].   ,,   ,,   mediæval pottery, more stratified than *b*[1].
b[3].   ,,   more washed out.

c.   Chalk Marl.
c'.   The Cambridge Greensand, about 1 foot in thickness, lying on an irregularly eroded surface of Gault.

d.   Gault.
d'.   Band of sandy clay with phosphate nodules similar to that above from which it is separated by about 1 foot of gault.

---

[1] It has been thought that some of these ridges might be due to flexures of the beds, but this explanation seems rarely applicable (*Mem. Geol. Surv. Explan. Quart. Sheet* 51, S.W. p. 38).

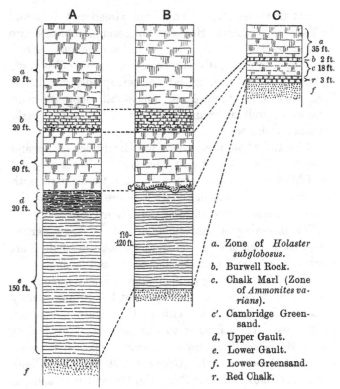

a. Zone of *Holaster*
    subglobosus.
b. Burwell Rock.
c. Chalk Marl (Zone
    of *Ammonites va-*
    rians).
c'. Cambridge Green-
    sand.
d. Upper Gault.
e. Lower Gault.
f. Lower Greensand.
r. Red Chalk.

FIG. 9. VERTICAL SECTIONS SHOWING DIMINUTION IN THICKNESS OF THE LOWER PART OF THE UPPER CRETACEOUS WHEN TRACED NORTH OF CAMBRIDGE.

Scale 100 ft. to 1 inch.

A. Hypothetical vertical section of the lower part of the Upper Cretaceous at Cambridge, showing the estimated thickness of the Gault *before* the total erosion of the Upper Gault and partial erosion of the Lower Gault at the time of the formation of the Cambridge Greensand.

B. Actual vertical section of the lower part of the Upper Cretaceous at Cambridge, showing the result of the erosion of 50 ft. of Gault at this period.

C. Actual vertical section of the lower part of the Upper Cretaceous at Hunstanton, showing diminution in the thickness of the beds due to thinning and not to erosion.

Occurrence. The Cambridge Greensand is found in East Bedfordshire, Hertfordshire, and Cambridgeshire along a tract of country over 50 miles in length. From Harlington in the south-west to a point a few miles north of Soham it has been traced uninterruptedly, and it has been worked for 'coprolites' almost all along its main outcrop as well as round the base of the large Chalk outliers to the west.

Over the whole area where the Cambridge Greensand is present the Upper Gault, the Upper Greensand, and the Chloritic Marl are absent, and the Chalk Marl with the Cambridge Greensand at its base rests directly on the eroded and irregular surface of the Lower Gault.

The incoming of these missing beds is first observed far south of Cambridge. Thus in tracing the Chalk Marl south-west of Harlington we find first a narrow area, some 10 miles wide, extending to the border of Buckinghamshire, over which the Upper Gault is present but is immediately succeeded by Chalk Marl with no Cambridge Greensand at its base. Then further to the south-west we find at Eaton Bray the first trace of true Upper Greensand, intercalated between the Upper Gault and the Chalk Marl. At Buckland near Tring the Upper Greensand has thoroughly set in, and from this point southward it increases in thickness and becomes the well-known calcareous free-stone and soft green sands of Oxfordshire and Berkshire.

The true Upper Greensand is thus seen not to pass laterally into the Cambridge Greensand but to be separated from the area where the latter occurs by a tract of country some 10 miles wide in which the Chalk Marl without a basal nodule bed rests directly on Upper Gault.

To the north of the Cambridge Greensand area it has been pointed out (p. 76) that though the whole of the

Gault is present yet it is most attenuated and altered in lithological character, and is directly overlaid by Chalk Marl of a hard rocky type.

The Cambridge Greensand is thus seen to be restricted in its occurrence, and to mark a local interruption in the stratigraphical sequence.

Mode of formation. A complete unconformity and sharp line of demarcation is thus proved to exist between the Gault and Cambridge Greensand. In fact the Cambridge Greensand which rests on a surface of erosion consists mainly of the débris of the upper portion of the Gault[1]. The matrix, which often is composed of Gault mud, the glauconitic grains which abound in the Upper Gault of adjacent areas, the black phosphatic nodules which are of the same type as those in the Upper Gault, and lastly but most particularly the derived fossils, most of which belong to Upper Gault species, distinctly indicate the origin of the material and the mode of formation of the Cambridge Greensand. Marine currents appear to have washed away the whole of the Upper Gault including the Vraconnian, and a portion of the Lower Gault over a certain well defined area. The materials were sorted and sifted and the greater part of the finer and softer clayey particles were removed while the larger and heavier sand grains and phosphatised fossils sank down to the bottom of the shallow sea.

There is on the other hand a complete and gradual passage up from the Cambridge Greensand into the overlying Chalk Marl: and not only is there this perfect stratigraphical continuity but the faunal characters show that the two must be closely associated (see p. 108).

[1] Jukes-Browne and Hill, *Q. J. G. S.* XLIII. (1887), p. 593 *et seq.*

100 THE CRETACEOUS BEDS.

Details of the constituents of the bed. *The
Matrix* consists of a fine chalky mud largely composed of
Foraminifera and minute fragments of various organisms[1].
In places it contains a large quantity of derived Gault
mud, as above mentioned. Grains of quartz, etc. also
occur.

The Glauconite grains are derived from the uppermost
beds of the Upper Gault where they are very abundant.
They are frequently the casts of Foraminifera[2]. On the
floor of existing seas we find internal casts of these
Protozoa being formed of the same substance.

The chemical composition of these grains is as
follows[3]:—

Water	10·80
Silica	51·09
Alumina	9·00
Iron protoxide	19·54
Magnesia	3·37
Lime	0·30
Soda	3·56
Potash	2·47
	100·13

The phosphatic nodules[4] vary in colour externally from
dark olive-black to brown, but when broken the interior
is found to be usually rather paler and browner or even
grey. A few of the nodules are of a pale buff colour and
probably do not contain so much phosphate as the darker
ones, and these lighter ones pass up into the main mass of
the Chalk Marl.

[1] Bonney, *Camb. Geol.* p. 32.
[2] Sollas, *Q. J. G. S.* xxviii. (1872), p. 397.
[3] Bonney, *op. cit.* p. 32.
[4] For general description of deposits of phosphate of lime see R. A.
Penrose, *Bull. U. S. Geol. Surv.* No. 46, 1888.

The average percentage of phosphate in the commoner dark nodules is from 56 to 58, but the amount varies considerably.

The nodules are mostly rounded lumps, varying in size from a mere grain to a fowl's egg, but are occasionally larger. Often they are of irregular shape, and sometimes they are tubular. In many cases it has been proved by microscopic examination that they are phosphatised sponges[1]. Others are the much worn and defaced casts of shells; others are lumps of phosphatised calcareous mud.

In chemical composition as well as in external appearance they are identical with the Upper Gault nodules of Bucks. and other counties. Dr Völcker[2] gives the following analyses of three samples:—

	I.	II.	III.
Moisture and organic matter	4·01	3·52	3·80
Lime	45·39	46·60	43·68
*Phosphoric acid	26·75	27·01	26·05
†Carbonic acid...	5·13	5·49	
Oxide of iron...	1·87	2·08	18·70
Alumina, magnesia, etc. ...	4·62	2·47	
Sulphuric acid, fluorine, etc.	6·01	6·79	
Insoluble siliceous matter ...	6·22	6·04	7·77
	100·00	100·00	100·00

* Equal to tribasic phosphate of lime	57·12	58·52	56·87
† Equal to carbonate of lime...	11·66	12·47	

These nodules are thus seen to be richer in phosphate of lime than those from the Lower Greensand or the Red Crag.

[1] Sollas, *Q. J. G. S.* xxix. (1873), pp. 63 and 76.
[2] Völcker, *Proc. Geol. Assoc.* vol. iii. (1872), p. 18.

Origin of the nodules. The organic origin of many of
the nodules is apparent at once from the fact that they are
recognisable fossils. Others, as above mentioned, require
the aid of the microscope to show their organic nature.
But apart from those of indisputable organic origin there
are many exhibiting no definite structure; and these may
represent organisms, which possessed no hard parts, or
may be only lumps of phosphatised mud.

The source of the phosphate is to be found in the
bones, tissues, and excreta of vertebrates, the bodies of
numerous invertebrates, marine plants growing in a
shallow sea, and possibly apatite in the detritus washed
down from the older rocks. The sea-water by containing carbonic acid, which is furnished by the decay of
organic matter, etc., would dissolve the phosphate and
hold it in solution. The mud of the sea-bottom would
thus be permeated by a solution of phosphate of lime.
But it is known that ammonia precipitates phosphate
of lime dissolved in carbonated water. And since ammonia is given off by decomposing animal matter, each
dead organism or portion of an organism would form
a centre round which the phosphate would be deposited.
Accordingly round these organic nuclei the phosphate of
lime would collect and ultimately form a nodule. A
combination, however, of more or less independent conditions appears necessary for the formation of phosphatic
nodules (see p. 74).

Many facts have been noticed showing the intimate
connection between the decay of animal matter and the
precipitation and concentration of phosphate of lime;
thus the roots of the shark's teeth are imbedded in a
mass of phosphate while the crown is free; the interior of
corals is filled with phosphate while the exterior is devoid

of it; the inner side of crabs' carapaces is coated with it while the outer side is bare and clean.

It has been noticed that the purer the phosphate in the interior of an organism the smaller is the aperture of the shell or test. Where the shell is open or has a large aperture the phosphate is mingled with grains of sand. These observations go to prove that the phosphate must have come from without in a state of solution and not as a soft paste, and that it cannot have been introduced after the formation of the deposit.

It must be remembered that the home of the Cambridge Greensand black nodules is the Upper Gault and that the process of phosphatisation went on in that deposit prior to its erosion. Only the paler coloured nodules which pass up into the Chalk Marl originated in the Cambridge Greensand.

The Erratics[1]. Far-travelled boulders in the Cambridge Greensand are of frequent occurrence and sometimes are so large as to cut out the nodule bed, but they generally range in diameter from about one inch to one foot. Amongst them are boulders of gneiss, granite, basalt, norite, quartz-porphyry, schist, greywacke, quartzite, sandstone, purple grit, and limestone. The blocks are sometimes rounded but often are subangular or even angular in shape. Pieces of a darkish-coloured amber[2] are not very rare, and pebbles of obsidian have been found.

Many of these boulders are covered with *Plicatula*

[1] Seeley, *Geol. Mag.* vol. III. 1866, p. 302; Jukes-Browne and Sollas, *Q. J. G. S.* XXIX. (1873), p. 11; Bonney, *Camb. Geol.* p. 33; *Mem. Geol. Surv. Explan. Quart. Sheet* 51, S.W. p. 25.

[2] Conwentz (*Natural Science*, vol. IX. (1896), No. 54, p. 102, has examined one specimen and states that it differs in both physical and chemical characters from succinite.

sigillina and an *Ostrea*, showing that they lay for some time on the sea-floor.

The home and mode of transportation of these blocks have been subjects of considerable discussion. Prof. Seeley held that they came from the south, referring them to France or some part of the East Anglian buried Palæozoic tract. But the Scotch or even Norwegian home of some of them seems in some cases extremely probable or even to be proved by their petrological characters. Moreover we have reason to believe that some of the rocks of which boulders are found do not occur to the south of our area.

With regard to the manner in which they were brought to Cambridgeshire some geologists have suggested icebergs, others have held that they may have been floated down entangled in masses of seaweed or in the roots of trees. Probably some of the blocks were brought in one way and some in the other, but the absence of ice striæ does not necessarily disprove the action of ice, for the stones resting on the surface of an iceberg would suffer no polishing, grinding, or striation during their sea-voyage or during the melting of the berg. The frequent occurrence of these blocks in the Cambridge Greensand and their almost complete absence from the beds of contemporaneous age in adjacent areas may well be due to the comparatively small depth of the sea on the floor of which the Cambridge Greensand accumulated. While the bergs would drift across the deeper seas in the neighbourhood without interruption or stoppage, they would ground in the shallows where the Upper Gault was being denuded, and as they overturned or melted, their load would be dropped and mingled with the sandy deposits on the sea-floor. Since the lithological characters of many of these boulders denote a northern origin we may suppose the existence of strong currents flowing southwards and carrying icebergs laden with rock fragments from their northern home, just as bergs now-a-days are floated southward by the Arctic currents into the Atlantic Ocean.

It is not however always easy to be sure that some of the recorded boulders are really from the Cambridge Greensand. Some are now known to have come out of superficial drift deposits, and others have been brought into the district as ballast or for paving purposes and been left on the surface of the ground. Unless a *Plicatula sigillina* or other characteristic Cambridge Greensand fossil is found adhering to the block, or unless the boulder has been found actually *in situ* it is wiser not to insist on its occurrence in the Cambridge Greensand.

Palæontology. Just as in the case of the Lower Greensand of this area, so in this bed the fossil contents fall naturally into two distinct groups which differ in their palæontological facies and in their physical condition.

There is (1) *the derivative fauna.* The fossils of this group are preserved in a black or very dark brown phosphate of lime of the same colour and appearance as the majority of the nodules: they are much rolled and water-worn. The bones of the vertebrates are generally rounded and worn, and the mollusca, etc. have generally lost their shells and are in the state of more or less battered casts: *Plicatulæ* and *Ostreæ* are often attached to their worn surfaces; and, finally, they occur at the very bottom of the deposit associated with nodules possessing similar physical characteristics.

Palæontologically this derived fauna has a very strong

Upper Gault facies, but we note the presence of some
Lower Gault forms, and also of other species with which
we are not familiar in any English formation, but which
we find in the 'Gault supérieur' or Vraconnian of the
Continent.

Thus three minor subdivisions may be made of this
derived fauna, containing respectively (1) the English
Lower Gault species; (2) the English Upper Gault
species; and (3) the Vraconnian or 'Gault supérieur'
species which come from a bed of Gault on a higher
horizon than any we have in England but which is not
equivalent to the true Upper Greensand of our southern
counties.

The very close relations of the derived fauna to that
of the Upper Gault are shown by the fact that 73 per
cent. of the English Upper Gault species occur in the
Cambridge Greensand. Only 39 per cent. of the Vraconnian and 34 per cent. of the Lower Gault forms are
found in it. Of true Upper Greensand species which
occur in our Cambridge bed there are only 28 per cent.,
and these are mostly bivalves with a wide range. 20 per
cent. of the Warminster Greensand fossils have been
recognised, but include none of the most characteristic
Warminster species[1].

It is, however, always the most common fossils which
give the true affinities of a fauna, and a list of those fossils
which may be picked up in any 'coprolite' pit near
Cambridge unmistakeably shows that it is the Upper
Gault fauna of England to which it bears the closest

---

[1] These percentages [*Q. J. G. S.* xxxi. (1875), p. 281] do not take any
account of the vertebrates since the state of preservation renders it often
doubtful whether we should ascribe a bone to the derived or to the
indigenous fauna.

resemblance, since all of the species, except four, in the following list are found in the Upper Gault of England. The four excepted fossils (*Hamites intermedius, Nautilus albensis, Pleurotomaria Rouxi, Rhynchonella sulcata*) occur in the Vraconnian of the Continent.

COMMON CAMBRIDGE GREENSAND FOSSILS[1].

REPTILIA.

Ichthyosaurus campylodon Carter

PISCES.

Edaphodon Sedgwicki Ag.
Lamna appendiculata Ag.

MOLLUSCA.

Cephalopoda.

Ammonites auritus Sow.
,, *mayorianus* D'Orb.
,, *raulinianus* D'Orb.
,, *rostratus* Sow.
,, *splendens* Sow.
Belemnites ultimus D'Orb. var.
Hamites intermedius Sow.
Nautilus albensis D'Orb.
,, *clementinus* D'Orb.

Gasteropoda.

Pleurotomaria Gibbsi Sow.
,, *Rhodani* Brong.
,, *Rouxi* D'Orb.
Solarium ornatum Sow.

Scaphopoda.

Dentalium decussatum Sow.

Lamellibranchiata.

Avicula gryphæoides Sow.
Exogyra rauliniana D'Orb.

[1] Jukes-Browne, *Q. J. G. S.* vol. XXXI. (1875), p. 280.

MOLLUSCA (*cont.*).

Lamellibranchiata (*cont.*).

Ostrea frons Park.
   „   *vesicularis* Sow.
Perna rauliniana D'Orb.
Plicatula pectinoides Sow.
Spondylus gibbosus D'Orb.

BRACHIOPODA.

Rhynchonella sulcata Park.
Terebratula biplicata Sow.

CRUSTACEA.

Palæocorystes Stokesi Mant.

ANNELIDA.

Serpula articulata Sow.

ACTINOZOA.

Trochocyathus angulatus Dunc.
   „   *conulus* Edw.

The fossils which were sifted out of the eroded beds were mixed in the sand with the remains of the animals that were flourishing in the sea whose waves and currents were carrying on their destructive work of wearing away the earlier formed clays.

(2) *The indigenous fauna.* We have now to consider the fauna which lived in the sea at the time this denudation was going on. From the fact that 50 per cent. of the species occurring in the true Chalk Marl of this and other areas are represented in this indigenous fauna it is evident that the palæontological characters of the bed support the conclusion which we drew from its stratigraphical relations, *i.e.* that it is an integral part of the Chalk Marl. Only about 16 per cent. of the indigenous species in the

Cambridge Greensand occur in the Upper Greensand of other districts, while there are about a dozen forms which have so far been found only in the Cambridge Greensand and appear to be peculiar to it.

The fossils of the indigenous fauna may be more or less easily separated from the derivative fauna by their mineral condition. The mineral condition is in fact the main test and distinction of the two faunas, for many of the species range down into lower beds or up into higher ones. The bones of the indigenous vertebrates are lighter in colour, more friable, and more perfect as a rule (at any rate before the artificial washing at the pits) than those of the derivative vertebrates. Their interstices also are commonly filled with the sandy marl of the bed itself. The calcareous shells of the Mollusca and Brachiopoda are always preserved and their interiors are filled either with an indurated sandy marl or with a semi-calcareous, semi-phosphatic material. These lighter coloured fossils together with the paler nodules range up into the overlying main mass of the Chalk Marl.

INDIGENOUS FAUNA OF THE CAMBRIDGE GREENSAND[1].

(N.B. The invertebrate species peculiar to the bed are marked thus *.)

REPTILIA.

> Acanthopholis eucercus Seeley
>   "   horridus Huxley ?
>   "   platypus Seeley
>   "   stereocercus Seeley
> Anoplosaurus curtonotus Seeley
>   "   major Seeley

---

[1] Jukes-Browne, *Q. J. G. S.* vol. xxxi. (1875), p. 280, *ib.* xxxiii. (1877), p. 485. *Mem. Geol. Surv. Explan. Quart. Sheet* 51, S.W. p. 149 (1881).

REPTILIA (*cont.*).

> *Cimoliosaurus Bernardi* Owen
> „   *planus* Owen
> *Emys* (?) *sphenognathus* Seeley
> *Ichthyosaurus campylodon* Carter
> *Macrurosaurus semnus* Seeley
> *Ornithocheirus Carteri* Seeley
> „   *Cuvieri* Seeley
> „   *Fittoni* Owen
> „   *macrorhinus* Seeley
> „   *platystomus* Seeley
> *Plesiosaurus cycnodeirus* Seeley
> „   *microdeirus* Seeley
> *Polyptychodon interruptus* Owen
> *Rhinochelys pulchriceps* Owen
> *Stereosaurus platyomus* Seeley
> „   *stenomus* Seeley
> *Testudo* (?) *cantabrigiensis* Seeley M.S.

PISCES.

> *Edaphodon Sedgwicki* Ag.
> *Enchodus halocyon* Ag.
> *Lamna appendiculata* Ag.
> *Notidanus microdon* Ag.
> *Protosphyræna ferox* Leidy

MOLLUSCA.

Cephalopoda.
> *Belemnites ultimus* D'Orb.

Lamellibranchiata.
> *Anomia transversa* Seeley
> *Avicula gryphæoides* Sow.
> *Lima globosa* Sow.
> * „   *ornata* Ether.
> *Ostrea cunabula* Seeley
> „   (*Exogyra*) *haliotoidea* Sow.
> „   *frons* Park.
> * „   *lagena* Seeley
> „   *vesicularis* Sow.

MOLLUSCA (*cont.*).

  Lamellibranchiata (*cont.*).
    Pecten orbicularis Sow.
    Plicatula inflata Sow.
    *   ,,   minuta* Seeley
      ,,   *sigillina* Woodw.
    Radiolites Moretoni Mant.
    Teredo amphisbœna Goldf.

BRACHIOPODA.

    Argiope megatrema Sow.
    Kingena lima Defr.
    Rhynchonella lineolata Phil.
      ,,   *mantelliana* Sow.
    Terebratula biplicata Sow.
      ,,   *semiglobosa* Sow.
      ,,   *sulcifera* Morris
    Terebratulina rigida var. ?
      ,,   *striata* Wahl. ?
    *   ,,   triangularis* Ether.[1]
    Thecidium Wetherelli Morris

ECHINODERMATA.

    Astrogonium sp.
    Cidaris Bowerbanki Forbes (spines)
      ,,   *Dixoni* Cotteau (spines)
    * ,,   gradata* Seeley
    * ,,   Sedgwicki* Seeley
      ,,   *vesiculosa* Goldf. (spines)
    Discoidea cylindrica Lam.
      ,,   *subucula* Klein
    Echinocyphus impressus Seeley
    Goniophorus lunatus Ag. var. *minutus* Seeley
    Holaster subglobosus Leske
    Pentacrinus Fittoni Austen
    Salenia Woodwardi Seeley

---

[1] This species used to be called *T. gracilis* and subsequently *T. striata* var. *triangularis*. It is distinct from the true *T. gracilis* of the Middle Chalk.

CRUSTACEA.

> *Palæga Carteri* Woodw.
> *Pollicipes arcuatum* Darw.
> „   *glaber* Sow.
> „   *unguis* Sow.

ANNELIDA.

> *Vermicularia umbonata* Mant. var.

ACTINOZOA.

> *Micrabacia coronula* Goldf.
> **Onchotrochus Carteri* Dunc.

HYDROZOA.

> *Parkeria* (large species)

PORIFERA.

> **Pharetrospongia Strahani* Sollas

FAUNA DERIVED FROM THE GAULT.

* indicates that the species occurs also in the English Gault.
† indicates that the species occurs also in the Gault supérieur (Vraconnian) of France and Switzerland.

REPTILIA.

> (?*)*Cetarthrosaurus Walkeri* Seeley
> *Chelone Jessoni* Lyd.
> *Cimoliosaurus Bernardi* Owen
>   *   „   *cantabrigiensis* Lyd.
>   „   *constrictus* Owen
>   „   *latispinus* Owen
> (?*)   „   *planus* Owen
> *Crocodilus cantabrigiensis* Seeley
>   „   *icenicus* Seeley
> *Eucercosaurus tanyspondylus* Seeley
> *Ichthyosaurus Bonneyi* Seeley
>   *   „   *campylodon* Carter
>   „   *Doughtyi* Seeley
>   „   *platymerus* Seeley

Reptilia (*cont.*).

Lytoloma cantabrigiensis Lyd.
Ophthalmosaurus (? *Baptanodon*) *cantabrigiensis* Lyd.
Ornithocheirus brachyrhinus Seeley
    „   *capito* Seeley
    „   *colorhinus* Seeley
    „   *crassidens* Seeley
    „   *Cuvieri* Bowerb.?
    „   *dentatus* Seeley
    „   *enchorhynchus* Seeley
    „   *eurygnathus* Seeley
    „   *machœrorhynchus* Seeley
    „   *microdon* Seeley
    „   *nasutus* Seeley
    „   *Oweni* Seeley
    „   *oxyrhinus* Seeley
    „   *platyrhinus* Seeley
    „   *polyodon* Seeley
    „   *scaphorhynchus* Seeley
    „   *Sedgwicki* Owen
            simus Owen
    ,   *tenuirostris* Seeley
    „   *xyphorhynchus* Seeley
Ornithostoma sp.
Patricosaurus merocratus Seeley
Plesiosaurus euryspondylus Seeley
    „   *ophiodeirus* Seeley
    „   *platydeirus* Seeley
    „   *pœcilospondylus* Seeley
(?*) *Polyptychodon interruptus* Owen
Protostega anglica Lyd.
Rhinochelys brachyrhina Lyd.
    „   *cantabrigiensis* Lyd.
    „   *elegans* Lyd.
    „   *Jessoni* Lyd.
    „   *macrorhina* Lyd.
    „   *pulchriceps* Owen
Stereosaurus cratynotus Seeley
*†*Syngonosaurus macrocercus* Seeley

REPTILIA (*cont.*).

    Trachodon cantabrigiensis Lyd.
    Trachydermochelys (? *Rhinochelys*) *phlyctænus* Seeley

AVES.

    Enaliornis Barretti Seeley
        ,,   *Sedgwicki* Seeley

PISCES [1].

    Anæmodus Carteri Sm. Woodw.
        ,,   *confertus* Sm. Woodw.
        ,,   *superbus* Sm. Woodw.
    Arthrodon crassus Sm. Woodw.
        ,,   *Jessoni* Sm. Woodw.
    Belonostomus sp.
    Cimolichthys (*Saurocephalus*) *striatus* Ag.
    Cælodus cantabrigiensis Sm. Woodw.
        ,,   *inæquidens* Sm. Woodw.
    Drepanophorus canaliculatus Eg.
    Edaphodon crassus Newton
    *   ,,   *laminosus* Newton
        ,,   *Reedi* Newton
        ,,   *Sedgwicki* Ag.
(?*)*Enchodus halocyon* Ag.
    **Hybodus* sp.
    **Ischyodus brevirostris* Ag.
        ,,   *latus* Newton
        ,,   *planus* Newton ?
    †*Lamna acuminata* Ag.
  *† ,,   *appendiculata* Ag.
  *† ,,   *gracilis* Ag.
     ,,   *plicatella* Reuss
  *† ,,   *subulata* Pict. and Camp.
    Lepidotus sp. (scales)
    Lophiostomus affinis Sm. Woodw.
    †*Oxyrhina macrorhiza* Pict. and Camp.
        ,,   *Mantelli* Ag. ?

---

[1] A. Smith Woodw. *Geol. Mag.* Dec. 3, vol. x. (1893), p. 489; and *ib.* Dec. 4, vol. xi. (1895), p. 207.

PISCES (*cont.*).

  *Pachyrhizodus glyphodus Blake ?
  Plethodus expansus Dixon
  *Portheus gaultinus Newton
  Protosphyræna brevirostris Sm. Woodw.
      „   depressa Sm. Woodw.
*   „   ferox Leidy
      „   Keepingi Sm. Woodw.
      „   ornata Sm. Woodw.
      „   tenuirostris Sm. Woodw.
  Ptychodus spectabilis Ag.
  Pycnodus cretaceus Ag.
    „   parallelus Eg. ?
  Sphenonchus sp.

MOLLUSCA.

Cephalopoda.

  *†Ammonites auritus Sow.
† „   „   var. renauxianus D'Orb.
   „   „   var. Salteri Sharpe
* „   cœlonotus Seeley
(?†) „   „   var. valbonnensis Heb.
   „   glossonotus Seeley
*† „   latidorsatus Mich. ?
*† „   planulatus Sow.
*† „   „   var. mayorianus D'Orb.
*† „   raulinianus D'Orb.
   „   „   var. tetragonus Seeley
(?†) „   rhamnonotus Seeley (=gardonicus Heb.)
   „   „   var. sexangulatus Seeley
*† „   rostratus Sow.
*† „   „   var. candollianus D'Orb. ?
*† „   „   var. inflatus Sow.
*† „   splendens Sow.
* „   „   var. cratus Seeley
   „   „   var. leptus Seeley
*† „   Studeri Pict. and Camp.
† „   timotheanus Mayor

MOLLUSCA (*cont.*).

Cephalopoda (*cont.*).

†*Ammonites vraconnensis* Pict.
    ,,   *Woodwardi* Seeley
**Ancyloceras tuberculatum* Sow.
*†*Anisoceras armatum* Sow.
*†   ,,   *saussureanum* Pict.
†*Baculites baculoides* D'Orb.
*†   ,,   *Gaudini* Pict. and Camp.
*†*Belemnites minimus* Lister
*†   ,,   *ultimus* D'Orb. (var.)
    Conoteuthis sp.
*†*Hamites intermedius* Sow.
*† ,, *virgulatus* Pict.
*†*Helicoceras robertianum* D'Orb.
    ,,   *4-tuberculatum* (? auct.)
*†*Nautilus albensis* D'Orb.
*(?†) ,, *arcuatus* Desh.
*†   ,, *clementinus* D'Orb.
*   ,, *inæqualis* Sow.
†   ,, *largilliertianus* D'Orb.
*   ,, *Montmollini* Pictet ?
*†*Scaphites hugardianus* D'Orb.
† ,,   ,,   var. *Meriani* Pict. and Camp.
* ,,   ,,   var. *simplex* Jukes-Browne
*†*Turrilites Bergeri* Brongn.
* ,, *elegans* D'Orb. ?
  ,, *emericianus* D'Orb. ?
*† ,, *hugardianus* D'Orb.
  ,, *nobilis* Jukes-Browne
† ,, *puzosianus* D'Orb.
  ,, *Wiesti* Sharpe var. *cantabrigiensis*

Gasteropoda.

Acmæa tenuicosta Desh.
  ,,   ,, var. *tenuistriata* Seeley
*†*Aporrhais carinata* Mant.
* ,, *elongata* Sow.
*† ,, *marginata* Sow.

MOLLUSCA (*cont.*).

  Gasteropoda (*cont.*).

   *†*Aporrhais Parkinsoni* Sow.

   *†*Avellana hugardiana* D'Orb.

   *†   „   *incrassata* Sow.

       „   *ventricosa* Seeley

       „   sp.

   **Brachystoma angulare* Seeley

   **Buccinum gaultinum* D'Orb.

   Cerithium sp.

   **Chemnitzia tenuistriata* Seeley

   †*Crepidula Cooksoniæ* Seeley

   *†   „   *gaultina* Buv.

   Funis elongatus Seeley

   Fusus quinquecostatus Seeley

  *(?†)   „   *Smithi* Sow.

       „   *tricostatus* Seeley

   Gibbula levistriata Seeley

   Hipponyx Dixoni Desh. ?

   Littorina crebricostata Seeley M.S.

   *†*Natica clementina* D'Orb.

   *†   „   *gaultina* D'Orb.

   *   „   *levistriata* Jukes-Browne

   †   „   *Rhodani* Pict. and Roux

   Nerinæa sp.

  *(?†)*Nerita cancellatus* Seeley (*indecisus* D'Orb.)

       „   *nodulosa* Jukes-Browne

       „   (*Neritopsis*) *scalaris* Seeley

   Ornithopus histochila Gard.

   *†   „   *retusus* Sow.

       „   „   var. *globulatus* Seeley

   †*Pleurotomaria allobrogensis* Pict. and Roux

   *†   „   *Gibbsi* Sow.

  (?*)†   „   *iteriana* Pict. and Camp. ?

       „   *Jukesi* Seeley

   †   „   *La Harpi* Pict. and Roux

  *(?†)   „   *lima* D'Orb.

   †   „   *regina* Pict. and Roux

   *†   „   *Rhodani* Brong.

MOLLUSCA (*cont.*).

Gasteropoda (*cont.*).

*†*Pleurotomaria Rouxi* D'Orb.

   ,, *semiconcava* Seeley

† ,, *vraconnensis* Pict. and Camp.

 Solarium Carteri Seeley

*† ,, *dentatum* Desh.

* ,, *granosum* D'Orb.

*† ,, *ornatum* Sow.

  ,, *planum* Seeley

† ,, *rochatianum* Pict. and Roux

  ,, *Sedgwicki* Seeley

 Stomatodon politus Seeley

 Tornatella pyrostoma Seeley

  ,, sp.

*†*Turbo pictetianus* D'Orb. (*nodosus* Seeley)

Scaphopoda.

 Dentalium decussatum Sow.

Lamellibranchiata.

*†*Arca hugardiana* D'Orb.

* ,, *nana* D'Orb.

*†*Avicula gryphæoides* Sow.

*†*Cardita tenuicosta* Sow. ?

*Cucullæa glabra Park.

*†*Exogyra conica* Sow.

* ,, ,, var. *plicata* Sow.

*† ,, *rauliniana* D'Orb.

†*Fimbria gaultina* Pict.

**Gervillia solenoides* Defr.

 Hinnites pectinatus Seeley

  ,, *Studeri* Pict. and Roux

  ,, *trilinearis* Seeley

*†*Inoceramus concentricus* Park.

*† ,, *sulcatus* Park. ?

†*Isoarca Agassizi* Pict. and Roux

*Leda solea D'Orb.

*†*Lima elongata* Sow.

* ,, *globosa* Sow.

  ,, *interlineata* Jukes-Browne

Mollusca (*cont.*).

   Lamellibranchiata (*cont.*).

      *†*Lima rauliniana* D'Orb.
      *Lucina tenera Sow.
      *†*Neithea quadricostata* Sow.
      *† ,,   *quinquecostata* Sow.
      *Nucula albensis D'Orb.
      *† ,,   *bivirgata* Sow.
      *† ,,   *ovata* Mant.
      ,,   *rhomboidea* Seeley
      ,,   *subelliptica* Seeley
      * ,,   *vibrayeana* D'Orb.
      *†*Ostrea frons* Park.
      *† ,,   *vesicularis* Sow.
    (?†)*Pecten aptiensis* D'Orb. var. *Barretti* Seeley
      *† ,,   *orbicularis* Sow.
      *† ,,   *raulinianus* D'Orb.
      ,,   *subacutus* D'Orb. ?
      *Perna lanceolata Gein.
      ,,   *oblonga* Seeley
      *† ,,   *rauliniana* D'Orb.
      ,,   *semielliptica* Seeley
      * ,,   *subspathulata* Reuss
    (?*)*Pholadomya decussata* Sow. var. *triangularis* Seeley
      *†*Plicatula pectinoides* Sow.
      *† ,,   *sigillina* Woodw.
      *†*Spondylus gibbosus* D'Orb.
      *†*Tellina phaseolina* Pict. and Camp.
      Teredo sp.

Brachiopoda.

      *†*Kingena lima* Defr.
      †*Rhynchonella dimidiata* Sow.
      ,,   ,,   var. *convexa*
    (?*)† ,,   *sulcata* Park.
      *†*Terebratula biplicata* Sow.
      ,,   ,,   var. *dutempleana* D'Orb.
      *† ,,   ,,   var. *obtusa* Sow.

CRUSTACEA.

Cyphonotus incertus Bell
**Diaulax carteriana* Bell
*Etyus Martini Mant.
     „   *similis* Bell
Eucorystes Carteri McCoy
Glyphæa Carteri Bell
*   „   *cretacea* McCoy
Hemioon Cuningtoni Bell
**Homolopsis Edwardi* Bell
**Hoploparia scabra* Bell
*   „   *sulcirostris* Bell
**Necrocarcinus Beechi* Deslong.
*   „   *tricarinatus* Bell
*   „   *Woodwardi* Bell
**Palæocorystes Stokesi* Mant.
Phlyctisoma granulatum Bell
     „   *tuberculatum* Bell
**Scalpellum arcuatum* Darw.
     „   *glabrum* Roem. ?
*   „   *læve* Sow.
*   „   *unguis* Sow.
Scillaridea cretacea Seeley M.S.
Squilla McCoyi Seeley M.S.
Xanthosia granulosa McCoy

ANNELIDA.

(?*)†*Serpula articulata* Sow.
   *   „   *plexus* Sow.

ECHINODERMATA.

*Cidaris gaultina Forbes
Galerites castaneus Brong. ?
*Hemiaster McCoyi Seeley
Hemipneustes sp.
†*Holaster lævis* Deluc.
**Pentacrinus Fittoni* Aust. ?
Pseudodiadema Barretti Woodw.
     „   *Carteri* Woodw.
     „   *fungoideum* Seeley

ECHINODERMATA (*cont.*).

Pseudodiadema intertuberculatum Seeley
,,   *inversum* Seeley
,,   *ornatum* Goldf.
,,   *scriptum* Seeley
,,   *variolare* Brong.
Salenia sp.

ACTINOZOA.

Isastræa sp.
*†*Trochocyathus angulatus* Dunc.
*†   ,,   *conulus* Edw.
*†   ,,   *harveyanus* Edw.

HYDROZOA.

Parkeria compressa Carter
,,   *nodosa* Carter ?
,,   *sphærica* Carter ?

PORIFERA.

Acanthophora Hartogi Soll.
*(?†)*Bonneyia bacilliformis* Soll.
,,   *cylindrica* Soll.
,,   *Jessoni* Soll.
,,   *scrobiculata* Soll.
,,   *verrongiformis* Soll.
Brachiolites tubulatus T. Sm.
Cephalites Benettiæ Mant.
,,   *capitatus* T. Sm.
,,   *compressus* T. Sm.
(?*)   ,,   *guttatus* T. Sm.
Eubrachus clausus Soll.
Hylospongia Bruni Soll.
,,   *calyx* Soll.
,,   *patera* Soll.
Pharetrospongia Strahani Soll.
Polyacantha Etheridgi Soll.
†*Retia costata* Soll.
*   ,,   *simplex* Soll.
Rhabdospongia communis Soll.

PORIFERA (*cont.*).

> *Ventriculites cavatus* T. Sm.
>   „   *mammillaris* T. Sm.
> *   „   *quincuncialis* T. Sm.
>   „   *texturatus* Goldf.

Exposures. The importation of phosphate of lime from America and Belgium and the small demand for the material owing to agricultural depression have caused most of the pits to be closed, and the majority of the pits mentioned in the Survey Memoir are now deserted or filled up.   Some are still open near Orwell and Horning-sea.   At the Mill Road Cement Works the nodules are dug out and washed when a certain area of the overlying Chalk Marl has been removed.

The relations of the Gault, Cambridge Greensand, and Chalk Marl are well seen in the large pits near Hauxton Mill, where the Chalk Marl is dug for cement.

The heaps of washed material usually yield on examination a number of fossils, but the best ones must be obtained from the workmen.   The more fragile specimens are broken in the process of washing and must be sought for and extracted from the bed *in situ.*

(b) THE UPPER PART OF THE ZONE OF *AMMONITES VARIANS.*

Characters and thickness. Having now considered the basement layer we turn to the main mass of the Chalk Marl[1]. This consists in Cambridgeshire principally of greyish calcareous marl, with occasional bands of harder blocky marl or 'clunch.' The lower part passes down

---

[1] The zonal subdivisions and nomenclature here adopted are those employed by Messrs W. Whitaker and Jukes-Browne in the " Geology of London " (*Mem. Geol. Surv.* 1889), vol. I. p. 58.

gradually into the Cambridge Greensand and is full of glauconitic grains.

The lithological characters of the Chalk Marl alter as it is traced north from Cambridge, the upper part first becoming harder and whiter, and ultimately the lower part also, till beyond Grimston in west Norfolk the whole division consists of hard chalk. North of Marham in the same county the lower marly portion with glauconitic grains gives place to a hard whitish limestone which at Hunstanton rests directly on the Red Chalk and is known as the "Sponge-bed" (Fig. 7, p. 84).

In addition to this lithological change the Chalk Marl in common with the other beds of the Lower Chalk suffers a considerable diminution in thickness as it is traced north. Throughout Cambridgeshire its thickness is everywhere about 60 feet, but beyond Stoke Ferry in Norfolk it is found to decrease, till at Hunstanton it is only 18 feet (Fig. 9, p. 97).

Occurrence. The Chalk Marl forms a low undulating strip of ground of varying breadth, lying for the most part on the east side of the valleys of the Cam and the Rhee.

The base of the formation has been traced by the 'coprolite' diggings in the Cambridge Greensand found all along its outcrop. It enters the county near Guilden Morden and then runs north-east in a somewhat irregular line past Harston to Cambridge. Its further course is along the east bank of the Cam to a little past Horningsea, beyond which it makes a sharp bend back southwards, and then is continued in a general north-easterly direction under Burwell Fen to Wicken and Soham, beyond which it is hidden by the fen deposits till we enter Norfolk.

The outcrop of the whole mass of Chalk Marl is

between one and two miles in breadth in the southern part of Cambridgeshire, but it increases in width near Shelford owing to the bay formed by the Granta valley. A promontory composed of higher beds stretches out to the west about half a mile south of Cavendish College (now the Homerton Training College) and narrows its outcrop, but north of this it again broadens considerably past Cherry Hinton and Fulbourn. A ridge of higher Chalk beds from the main mass of Chalk in the east juts out as far as Stow-cum-Quy and separates off another bay to the north near Bottisham. Beyond this latter bay the outcrop narrows and runs past Reach, Fordham, and Burwell to sink below the fens till we reach Norfolk. Cherry Hinton, Teversham, Fen Ditton, Horningsea, Stow-cum-Quy, Wilbraham Fen, Bottisham, Swaffham Bulbeck, Long Meadow, Reach and Burwell Fen are all situated on this principal line of outcrop of Chalk Marl.

Numerous outliers occur to the west of this main mass. The largest one of these is between Barrington and Eversden, and has its high central portion formed of higher beds of Chalk capped by Boulder Clay. Its boundary, roughly speaking, runs by Orwell and Wimpole, then north to Kingston, and then south-east through Eversden to Haslingfield, where it bends south to Barrington and Harston. This mass is for all practical purposes an outlier though it may be connected with the main mass of Chalk to the east by a narrow neck of Chalk Marl below the alluvium of Harston. A smaller outlier to the south lies near Arrington.

The next most important outlier is the Grantchester-Coton mass. The boundary of this passes from Grantchester to Barton, thence to Hardwick and thence east to Coton. South of this village and on the east side of the outlier there is a huge bay running into it and nearly dividing it into two parts. The western part of this outlier is concealed beneath Boulder Clay.

A small outlier lies close to the northern cape of the large bay in the Coton outlier, and forms the ridge which extends eastwards from the Observatory to Castle Hill and northwards on the road to Girton.

There are several less important outlying patches of Chalk Marl, as shown on the one-inch Geological Survey map, near Orwell, Comberton, Haslingfield and Chesterton.

Exposures. The lowest beds are seen in most of the 'coprolite' diggings in the Cambridge Greensand, as at Hauxton, Barrington, and the Mill Road Cement Works.

The upper part of the Chalk Marl is seen in the northern pit at Burwell, underlying the Burwell Rock (see p. 127). The Madingley pit, S.E. of Madingley village, by the roadside, is too much overgrown now to allow the section to be examined.

Near Shepreth, close to the Great Northern Railway, are several large cement pits in the Chalk Marl.

Economics. The Chalk Marl is largely dug for cement. Pits are open and large works erected on the Mill Road, Cambridge, also near Shepreth, and at other places between Cambridge and Royston.

From the base of the Chalk Marl springs are thrown out, the water being held up by the Gault below. From the fact that the water issues from the basal phosphatic nodule or 'coprolite' bed these springs are sometimes locally called the "fossil springs."

Agriculturally the Chalk Marl forms a rich tract; the soil is entirely under cultivation and produces rich corn-crops.

FOSSILS OF THE CHALK MARL.

(ZONE OF *Ammonites varians.*)

PISCES.

Lamna appendiculata Ag.

MOLLUSCA.

Cephalopoda.

Ammonites varians Sow.

Lamellibranchiata.

Inoceramus latus var. *reachensis* Ether.
Lima globosa Sow.
Ostrea vesicularis Lam.
Pecten orbicularis Sow.
Plicatula inflata Sow.

BRACHIOPODA.

Kingena lima Defr.
Rhynchonella grasiana D'Orb.
    „   *mantelliana* Sow.
    „   *Martini* Mant.
Terebratula biplicata Sow.
    „   *semiglobosa* Sow.
Terebratulina gracilis var. *nodulosa* Ether.
    „   *striata* Wahl.
    „   *triangularis* Ether.

ECHINODERMATA.

Cidaris dissimilis (spines) Forbes ?
Discoidea subucula Klein
Holaster subglobosus Leske ?

ANNELIDA.

Serpula annulata Reuss
Vermicularia umbonata Sow.

(2) THE ZONE OF *HOLASTER SUBGLOBOSUS*.

(a) THE BURWELL ROCK OR TOTTERNHOE STONE.

General remarks. This well-marked bed forms the base of the Zone of *Holaster subglobosus* throughout eastern and south-eastern England. In Cambridgeshire the local name "Burwell Rock" is applied to it, and as a water-bearing stratum and source of a local building stone it is of considerable importance.

Characters and thickness. The bed consists of a light brown or buff sandy chalk, hard and gritty at the base and compact and homogeneous above. The roughness and gritty texture of the rock are largely due to the abundance of small broken fragments of *Inoceramus*-shells. Cal-careo-phosphatic nodules are scattered through the bed but are chiefly collected in the basement layer which is full also of organic remains. This basement layer—locally known as 'brassil'—is composed of a hard grey gritty stone crowded with the calcareo-phosphatic nodules which have an apple-green coat but are of a light brownish-yellow colour inside. They vary from the size of a pea to that of a potato. The overlying beds consist of a compact grey rock which is easy to dress, and it is this stone which is and has been for centuries quarried for building purposes.

Round Cambridge the thickness of the Burwell Rock is about 15 feet, as at Cherry Hinton. At Burwell it is about 20 feet thick, and of this thickness the basement layer occupies from six inches to a foot. At Stoke Ferry the whole bed has decreased to 5 feet in thickness, and at Dersingham to 2½ feet, while at Hunstanton it is only 2 feet thick.

The fine quartz sand which is one of the constituents of the rock at Totternhoe and is also present in Cambridgeshire is almost absent at Hunstanton.

The persistent and marked characters of this rock are most useful in tracing the Chalk across country. But it is seen that in common with the underlying beds it thins towards the north and shows a diminution in the amount of mechanical sediment.

Occurrence. The Totternhoe Stone forms an important horizon in Buckinghamshire, Bedfordshire and Hertfordshire, and its mode of occurrence in these counties has been described by Mr Whitaker (*Mem. Geol. Surv.* vol. IV. 1872).

From Ashwell it has been traced past Shepreth, Harston, and Shelford to Cherry Hinton. At various points along its course it throws out springs from its base, the water being held up by the argillaceous marls below. The Nine Wells at Shelford and Shardelow's Well near Fulbourn are instances. These springs are of great assistance in following its outcrop. At Cherry Hinton the topmost four feet of the Stone is exposed in the lower part of the quarries. Near Fulbourn the water from it forms the main supply of the Wilbraham river. Further to the north-east the Lower Chalk is disturbed and two or more anticlinal axes exist which bring up the Chalk Marl on each side of Burwell. At Burwell itself in the three quarries the Burwell Rock is well seen. North of Burwell the outcrop runs to Fordham and Isleham, and beyond it is more or less hidden beneath the fens till we enter Norfolk. In the outliers to the west this rock is seen in several quarries, as at Haslingfield.

Economics. Near Swaffham Prior and Burwell the bed is burnt for lime.

The hard basement layer or 'brassil' is broken up and used for road-metal in the fens.

The compact and easily dressed rock has long been quarried for building purposes at Burwell, and was formerly procured at Cherry Hinton. It is especially suitable for internal decoration, being quite durable when not exposed to the weather.

The bed is an important water-bearing stratum, more from the occurrence of numerous joints and fissures than from its porous character.

The water which accumulates in it by percolating through the overlying permeable Chalk is thrown out as springs along its outcrop as above mentioned. Owing to the occurrence of the synclinal beneath the Gog Magog Hills which affects the Burwell Rock in common with the other beds an excessive amount of water collects in the fissures which traverse it. Where the trough emerges on the surface of the ground, as at Cherry Hinton, an unusually strong spring is found.

Palæontology. The fossils of this bed are abundant, especially in the basement layer. In the following list the commonest forms are marked with an asterisk.

PISCES.

> *Beryx* sp.
> Corax sp.
> *Dercetis* sp.
> Lamna appendiculata Ag.
> Notidanus sp.
> Ptychodus sp.

Mollusca.

Cephalopoda.

Ammonites lewesiensis Mant. ?

    „   *Mantelli* Sow. var.

    „   *navicularis* Mant.

*   „   *rhotomagensis* Brong.

    „   „   var. *cenomanensis* D'Arch.

*   „   *varians* Sow.

    „   „   var. *Coupei* Brong.

Belemnitella plena Blainv. var. ?

Nautilus deslongchampsianus D'Orb.

    „   *elegans* Sow.

    „   *pseudoelegans* D'Orb.

    „   *reflectus* Seeley M.S.

Scaphites æqualis Sow.

Trigonellites sp.

Turrilites costatus Lam.

    „   *scheuchzerianus* Bosc.

    „   *tuberculatus* Bosc.

Gasteropoda.

Aporrhais sp.

Cerithium ornatissimum Desh.

Pleurotomaria perspectiva Mant.

    „   sp.

Scalaria fasciata Ether.

Solarium dentatum Desh. ?

Trochus sp.

Turbo sp.

Scaphopoda.

Dentalium majus

Lamellibranchiata.

Anomia papyracea D'Orb. var. *burwellensis* Ether.

Avicula dubia Ether.

    „   *filata* Ether.

    „   *gryphæoides* Sow.

Inoceramus convexus Ether.

    „   „   var. *quadratus* Ether.

    „   *latus* Mant. var. *reachensis* Ether.

Mollusca (*cont.*).

  Lamellibranchiata (*cont.*).

    Lima aspera Mant.

      „ *echinata* Ether.

     * „ *globosa* Sow.

    Neithea quinquecostatus Sow.

    Ostrea acutirostris Nilss.

      „ *curvirostris* Nilss. var. *inflexa* Ether.

      „ *frons* Park.

      „ (*Exogyra*) *haliotoidea* Sow.

      „ *vesicularis* Lam.

    Pecten Beaveri Sow.

      „ *elongatus* Lam. ?

     * „ *fissicosta* Ether.

     * „ *orbicularis* Sow.

    Pholadomya decussata Phil.

    Pinna tegulata Ether.

    Plicatula inflata Sow.

    Spondylus striatus Sow.

    Teredo amphisbœna Goldf.

Brachiopoda.

    Kingena lima Defr.

    Rhynchonella grasiana D'Orb.

     *   „   *mantelliana* Sow.

         „   *Martini* Mant.

    Terebratula biplicata Sow.

     *   „   *semiglobosa* Sow.

         „   *squamosa* Mant.

    Terebratulina gracilis Schl. var. *nodulosa* Ether.

         „   *striata* Wahl.

Echinodermata.

    Cidaris dissimilis (spine) Forbes ?

    Cyphosoma sp. ?

    Discoidea subucula Klein

    Hemiaster Morrisi Forbes

    Holaster subglobosus Leske

    Pentacrinus Fittoni Aust.

    Pseudodiadema sp.

CRUSTACEA.

> *Enoploclytia brevimana* McCoy
> „   *Imajei* Mant.
> *Glyphæa cretacea* McCoy
> *Necrocarcinus Woodwardi* Bell
> *Pollicipes glaber* Rom.

ANNELIDA.

> *Serpula antiquata* Sow.
> „   *rustica* Sow.
> *Vermicularia umbonata* Sow.

ACTINOZOA.

> *Cælosmilia* sp. ?
> *Micrabacia coronula* Goldf.
> *Onchotrochus serpentinus* Sow.

(b) THE BEDS BELONGING TO THE ZONE OF *HOLASTER SUBGLOBOSUS* ABOVE THE BURWELL ROCK.

Characters and thickness. These beds are about 80 feet thick at Cherry Hinton; the lower 30 feet or so consist of ordinary grey chalk; the upper part of hard white chalk. Both parts possess remarkable curved divisional lines, which appear to be entirely due to a kind of curvitabular jointing and not to lenticular bedding. The irregularity of the bedding is characteristic of the whole zone. Messrs Jukes-Browne and Penning express the following opinion[1]: "It is a matter of some doubt whether this structure can be termed bedding in the true sense of the word. We question whether it is any proof of current action, and would suggest that the curved surfaces may be due to the contraction of a homogeneous mass in the absence of any definite bedding planes. We

[1] *Mem. Geol. Surv. Explan. Quart. Sheet* 51, S.W. p. 51.

believe, in fact, that they are lines of jointing rather than of bedding."

There are nodule beds at Litlington and Swaffham Bulbeck in the upper part of this zone, but they are not continuous and are absent at Cherry Hinton.

Local erosion of the upper part appears to have taken place before the deposition of the Belemnite Marls, as at Chalkshire near Wendover[1], but the zone is at its thickest near Cambridge.

At Hunstanton the total thickness of the beds is only 35 feet, for they partake in the general thinning of the Lower Chalk when followed northwards.

Occurrence. The outcrop of these beds is defined by that of the underlying Burwell Rock and that of the overlying Belemnite Marls and Melbourn Rock. It runs from Hitchin a little north of Royston, past Melbourn, Foxton, Newton and Shelford to Cherry Hinton, occupying an area which averages a mile and a half to two miles in breadth. It passes then eastwards to Fulbourn, forming the steep slopes down to the village and the high ground on the south-east border of Fulbourn Fen. Thence it stretches to Great Wilbraham and passes Swaffham Prior and Burwell to bend sharply south-eastward towards Exning and Newmarket. North of the latter place it runs to the west of Mildenhall and passes into Suffolk and Norfolk[2], but is considerably hidden beneath the fen deposits.

The large outliers of the Upper Cretaceous which lie

[1] W. Hill and Jukes-Browne, "The Melbourn Rock," etc., *Q. J. G. S.* vol. XLII. (1886), p. 227.

[2] W. Hill and Jukes-Browne, "On the Lower Part of the Upper Cretaceous in West Suffolk and Norfolk," *Q. J. G. S.* vol. XLIII. (1887), p. 563.

on the Gault plain to the west of the Cam are capped by beds belonging to this horizon.

Exposures. The Cherry Hinton chalk-pit offers one of the best and most accessible sections in the neighbourhood. The zone is here about 80 feet thick. The western face of the quarry shows the greyish chalk—about 30 feet thick—forming the lower part of this zone and resting on the Burwell Rock. Above this greyish chalk the thicker and harder white chalk is seen in the eastern and higher parts of the quarry. The curved jointing is very well displayed in this pit, and fossils in the lower beds are fairly common. The clunch pit at Newton appears to be opened in the upper beds.

The lower beds with their characteristic fossils are well seen in the pits at Burwell.

Palæontology. The following fossils have been found in the Cambridgeshire area in this zone, but the fauna is very scanty:—

REPTILIA.

    Dolichosaurus longicollis Owen
    Saurospondylus dissimilis Seeley (? from this zone)

MOLLUSCA.

    Pecten fissicosta Ether.
      „   *orbicularis* Sow.

BRACHIOPODA.

    Rhynchonella mantelliana Sow.
        „   *Martini* Mant.
    Terebratula semiglobosa Sow.
    Terebratulina striata Wahl.

ECHINODERMATA.

    Holaster subglobosus Leske

(3) THE ZONE OF *BELEMNITELLA PLENA*.

General remarks. It is only since the year 1886 that this zone has been definitely traced through this area[1]. Formerly it was not separated from the overlying Melbourn Rock, but was grouped with it at the base of the Middle Chalk, and this plan was followed by the Geological Survey in 1881. But now these marls containing *Belemnitella plena* must be regarded as forming the top of the Lower Chalk. They are commonly known as the "Belemnite Marls."

Characters and thickness. In Cambridgeshire this zone is composed of two principal layers of marl separated by a hard rocky band.

The lower layer of marl is shaly, of a greenish-grey colour, and in places passes down gradually into the chalk beneath, but elsewhere rests on an uneven surface of tough dull white chalk which has different microscopical characters from those of the chalk elsewhere underlying the marls. This change in the relation of the basal layer to the beds below is held by Messrs Hill and Jukes-Browne to be due to contemporaneous erosion[2].

The lower layer of marl averages 1 to 1½ feet in thickness throughout Cambridgeshire. It contains in places (*e.g.* Royston) pieces of unaltered white chalk similar in structure to the Lower Chalk into which the marl passes. Occasionally these fragments are so numerous as to give a mottled appearance to the marl.

---

[1] W. Hill and A. J. Jukes-Browne, "The Melbourn Rock," etc., *Q. J. G. S.* vol. XLII. (1886), p. 216.

[2] *op. cit.* p. 225.

The middle rocky band consists of hard white smooth chalk passing occasionally into a peculiar nodular 'marbled' rock, which is found to be composed of "more or less rounded fragments of chalk from the size of an egg to that of a pin's head, set in a greyish matrix, composed largely of shell-fragments," and it is held by Messrs Jukes-Browne and Hill to be the result of the breaking up of a bed of semi-consolidated chalk by the action of a gentle current. Some of these characters are however probably due to concretionary action. This band is about 2 feet in thickness.

The upper layer of marl varies from 3 to 5 feet in thickness and is of a greyish-buff colour with a harsh gritty texture.

Occurrence. To the south of Cambridge at Shelford, Harston, Foxton, Melbourn, Royston, Litlington, and Ashwell this zone has been discovered in several pits, and details of these sections are given by Messrs Hill and Jukes-Browne (*op. cit.*). At Cherry Hinton a narrow band of yellowish gritty laminated marl represents it, but further to the north-east its persistence is largely conjectural, owing to lack of exposures. The overlying Melbourn Rock forms a more or less distinct surface feature, and the Belemnite Marls are presumably present beneath it all the way to Swaffham Bulbeck, where they are seen in two pits east of the village[1].

In Norfolk the muddy marly element is gradually eliminated as the beds are followed northwards, and in the absence of fossil evidence the horizon cannot be distinguished from the contiguous beds of chalk.

---

[1] *Mem. Geol. Surv. Explan. Quart. Sheet* 51, S.W. p. 59.

Exposures. In the pit on Steeple Hill at Shelford there is a fairly good exposure, but it is considerably weathered.

The lime-kiln pit at Melbourn gives the following section[1]:—

		ft.
	Top soil, rubble and much broken chalk ...	8
	Bedded white chalk	4
Melbourn	Thin bedded yellowish chalk, rather hard ...	4
Rock	Hard rough nodular rock	3½
Zone of	Soft laminated marly bands enclosing harder greyish marly chalk	1
Bel. plena	Grey marly chalk	2
	Hard white rocky chalk with smooth fracture	2
	Grey marly chalk	1½
	White chalk	2

In a pit by the roadside a mile N.W. of Royston the 'marbled rock' is seen resting on grey marly chalk and overlaid by buff-coloured thin beds of marl, etc.

Palæontology. The following fossils are recorded by Messrs Jukes-Browne and Hill from this zone in Cambridgeshire :—

MOLLUSCA.
  Belemnitella plena Blainv.
  Ostrea vesicularis Lam. var. *Baylei*
BRACHIOPODA.
  Rhynchonella plicatilis Sow.
  Terebratula biplicata Broc.
    „   *semiglobosa* Sow.

N.B. It should be noted that the total thickness of the Lower Chalk diminishes from 170 feet at Newmarket to 55 feet at Hunstanton. The changes in the individual beds have been mentioned above.

[1] W. Hill and Jukes-Browne, *op. cit.* p. 221.

THE MIDDLE CHALK.

(1) THE ZONE OF *RHYNCHONELLA CUVIERI.*

General remarks. This is a group of beds agreeing
closely with that designated by Dr Barrois as the "Zone
of *Inoceramus labiatus,*" but Messrs Jukes-Browne and
Penning in the Survey Memoir on Sheet 51, S.W., and
subsequent writers on the Chalk of this area, prefer to call
it the "Zone of *Rhynchonella Cuvieri*" on account of the
abundance of that species of *Rhynchonella,* and they include
the Melbourn Rock at its base.

(a) THE MELBOURN ROCK.

The Melbourn Rock consists of a persistent and conspicuous group of hard rocky beds at the base of the
Middle Chalk and of the Zone of *Rhynchonella Cuvieri.*
Since the publication of the Survey Memoir its lower beds
have been cut off to form the new subdivision, called the
Zone of *Belemnitella plena,* above described. It should be
borne in mind that the Melbourn Rock is a lithological
horizon rather than a bed with special palæontological
characteristics, and that its organic remains show its
close connection with the beds immediately overlying it.

Characters and thickness. The lowest bed of the
Melbourn Rock, immediately overlying the Zone of *Bel.
plena,* is a hard nodular rock, 3 or 4 feet thick, consisting
of greenish-grey chalk crowded with small white nodules.
The upper beds are of yellowish chalk with thin partings
of marl and layers of small nodules from 6 to 18 inches

apart. A few nodules are also scattered through the mass. A band of smooth rock sometimes occurs at a height of 9 or 10 feet from the base.

The matrix and nodules are of the same general composition; but the heavier particles, consisting chiefly of shell-fragments, are more abundant in the matrix, which appears to consist of the washed and sorted material composing the nodules. The total thickness of the Melbourn Rock in Cambridgeshire varies from 6 to 10 feet.

Occurrence. The Melbourn Rock is exposed at the localities mentioned above in describing the Belemnite Marls.

The beds are generally found capping the escarpment of the Lower Chalk, and give rise to a more or less marked feature in the landscape, particularly in the southern and central parts of the area. Outside Cambridgeshire, in the adjoining counties of Norfolk, Suffolk, Herts., etc. the Melbourn Rock has been traced as a persistent horizon with substantially the same characters, and the "Grit bed[1]" on the Kentish coast at the base of the Zone of *Rhynchonella Cuvieri* exactly corresponds to it.

Palæontology. The following fossils have been found in the Melbourn Rock in Cambridgeshire:—

MOLLUSCA.
 Inoceramus sp.
BRACHIOPODA.
 Rhynchonella Cuvieri D'Orb.
 Terebratulina striata Wahl.
ECHINODERMATA.
 Cidaris sp. (spines)

[1] F. G. H. Price, *Q. J. G. S.* vol. XXXIII. (1877), p. 431, and W. Hill, *Q. J. G. S.* vol. XLII. (1886), p. 232.

(b) THE BEDS BELONGING TO THE ZONE OF *RHYN-CHONELLA CUVIERI* ABOVE THE MELBOURN ROCK.

Characters and thickness. Bedded white chalk with occasional flints and thin marly layers compose the remainder of the zone, and have a thickness of about 60 to 70 feet.

Occurrence. The area occupied by the outcrop of these beds lies between the two well-marked bands of hard rock—the Melbourn Rock below and the Chalk Rock above. Royston, Triplow, and Pampisford lie upon them, and they form the upper part of the high chalk ridge and promontory on the north side of the valley of the Lin extending from Linton to Cherry Hinton. Together with the overlying "Wandlebury Beds" and the Zone of *Holaster planus* they form the northern and western slopes of the Gog Magog Hills. Newmarket stands on beds of this and the succeeding horizon.

Exposures. There are few good sections of these beds. By Stanmoor Hall west of Whittlesford there is a small pit where beneath a few feet of river-gravel 8 or 10 feet of rotten lumpy chalk are seen containing the zone fossil. A pit in a wood near Babraham has furnished many fossils from the upper part of this zone. The rock is exposed also at the top of the chalk-pit near the reservoir above the great Cherry Hinton quarry.

Palæontology. The following fossils have been found in these beds of the Zone of *Rhynchonella Cuvieri* in this area :—

PISCES.
  Ptychodus decurrens Ag. ?

MOLLUSCA.

Lamellibranchiata.

Inoceramus labiatus Brong. (=*mytiloides* Mant.)
,,   ,,   var. *problematicus* D'Orb.

BRACHIOPODA.

Rhynchonella Cuvieri D'Orb.
,,   ,,   var. (or young)
,,   *mantelliana* Sow.
,,   *Martini* Mant.
Terebratula semiglobosa Sow.
Terebratulina gracilis Schloth. var. *lata* Ether.

ANNELIDA.

Serpula antiquata Sow. ?

ECHINODERMATA.

Cidaris dissimilis Forbes (spines)
Echinoconus globulus Desor.
,,   *subrotundus* Mant.

(2) THE ZONE OF *TEREBRATULINA GRACILIS.*

General remarks. The strata composing this zone in Cambridgeshire have been termed the "Wandlebury Beds." They are composed of hard beds of chalk at the base, and cause a marked local rise in the ground.

Formerly the beds which are now separated off as the Zone of *Holaster planus* were included as the upper division of the Zone of *Terebratulina gracilis* (*Geological Survey Memoir Explan. Quart. Sheet* 51, S.W. p. 62).

Characters and thickness. The zone consists of hard massive white chalk resembling the Chalk Rock. *Rhynchonella Cuvieri* and *Inoceramus labiatus* var. *problematicus* are fairly common fossils in it.

The total thickness of the zone is about 100 feet.

Occurrence and exposures. There is a great lack of exposures of these beds in this area. Near Royston the lower part of the zone is seen in some quarries, and the shape of the ground near Pampisford Hall leads one to suppose that it occurs here also. It crosses the railway near Babraham and appears in a small pit south of Little Abington Grange. It is exposed again near Fulbourn, Worsted Lodge, and Mutlow Hill.

Palæontology. The following fossils from this zone have been recorded in our area:—

MOLLUSCA.
    Lamellibranchiata.
        Inoceramus Brongniarti Sow.
           „   *labiatus* Brong.
           „   „   var. *problematicus* D'Orb.

MOLLUSCA (*cont.*).

  Lamellibranchiata (*cont.*).

    Lima striata Sow.
    Ostrea vesicularis Lam.
    Pecten Beaveri Sow.

BRACHIOPODA.

    Rhynchonella Cuvieri D'Orb.
          „   „   var. (or young)
          „   *plicatilis* Sow.
    Terebratula semiglobosa Sow.
    Terebratulina gracilis Schloth. var. *lata* Ether.
         „   *striata* Wahl.

ECHINODERMATA.

    Cidaris dissimilis Forbes (spines)
    Discoidea Dixoni Forbes
    Echinoconus subrotundus Mant.

ACTINOZOA.

    Parasmilia sp.

(3) THE ZONE OF *HOLASTER PLANUS.*

General remarks. These beds, once assigned to the underlying zone, are now[1] recognised as the distinct zone established by Barrois under the name of the Zone of *Holaster planus.*

Characters and thickness. The zone, which is about 50 feet thick, consists of soft white homogeneous chalk with occasional layers of marl and flints. Silicified sponges are generally common.

Occurrence and exposures. The hill-slopes below the outcrop of the Chalk Rock are composed of these beds, but there are very few exposures. The zone however may be seen in a quarry north of Linton, where the zone fossil occurs; and in some pits near Westley Waterless; and lastly in a quarry a little N.E. of Dullingham Station.

Palæontology. The following fossils are found in this zone in the Cambridge area :—

MOLLUSCA.

> *Inoceramus Brongniarti* Sow.
> *Lima striata* Sow.
> *Ostrea vesicularis* Lam.
> *Pecten Beaveri* Sow.
> *Spondylus spinosus* Sow.

BRACHIOPODA.

> *Rhynchonella Cuvieri* D'Orb.
>   ,,   *plicatilis* Sow.
>   ,,   *reedensis* Ether.
> *Terebratula semiglobosa* Sow.

[1] *Mem. Geol. Surv.* "Geology of London" (1889), p. 58.

ECHINODERMATA.

> *Cidaris sceptrifera* Mant.
> *Cyphosoma radiatum* Sorig.
> *Holaster planus* Mant.
> *Micraster breviporus* Ag.
> „ *cor-bovis* Forbes ?

PORIFERA.

> *Ventriculites impressus* Smith
> „ *mammillaris* Smith

(4) THE CHALK ROCK OR ZONE OF *HETEROCERAS REUSSIANUM*.

General remarks. The Chalk Rock forms the top of the Middle Chalk, separating it from the Upper Chalk. Its outcrop near Wardington Bottom, near Royston, has revealed a remarkable flexure in the beds (see next page)[1].

Characters and thickness. The Chalk Rock is a hard crystalline bed of chalk, frequently of a yellow colour and broken into lumps enclosed in a marly matrix. There is sometimes a softer layer of marly chalk in the middle of the hard crystalline bed, dividing it into two parts; the lower part passes down gradually into the white chalk below, but the upper part is sharply defined from the Upper Chalk. The total thickness varies from 2 to 14 feet. Thus, in a chalk pit a mile north of Reed it consists of three hard beds of chalk amounting to 14 feet in thickness; but near Westley Waterless it is represented by a layer of crystalline yellowish chalk only two to three feet thick.

---

[1] *Mem. Geol. Surv.*, Explan. Sheet 47, pp. 7—11, figs. 1 and 2; *Mem. Geol. Surv.*, Explan. Quart. Sheet 51 S.W., p. 67.

R. 10

From a study of the mollusca Mr H. Woods considers that the Chalk Rock was laid down between the depths of about 100 and 500 fathoms[1].

Occurrence and exposures.   South-east of Royston the outcrop of the Chalk Rock describes a curve owing to a line of flexure.   The pits near Reed, Barkway, Heydon, and Barley are situated on this line, and afford good sections of the Chalk Rock[2].   The line of flexure encloses the head of Wardington Bottom, and the beds dip inwards to the north and north-west at high angles, instead of at the usual low angles to the south or south-east.   The disturbance causing this depression and reversal of dip appears to have been quite local, but it is interesting from its influence on the surface contours, though it was formed prior to the deposition of the Boulder Clay.

A pit S.S.E. of Great Chesterford gives another exposure of the Chalk Rock; and from thence the outcrop runs past Abington to Balsham by Conger's Well and Linnet's Hall to Westley Waterless, where in a pit $\frac{3}{4}$ mile S.W. of the village the Chalk Rock is seen to be composed of only two or three feet of much broken-up yellowish crystalline chalk.   Many fossils have been obtained from this bed in the pits at Underwood Hall, Dullingham, and near Stetchworth.

Palæontology.   The following fossils[3] from the Chalk Rock have been found in this area:—

[1] H. Woods, *Q. J. G. S.* vol. LIII. (1897), p. 402.
[2] *Mem. Geol. Surv.*, Explan. Sheet 47, pp. 7 and 8, figs. 1 and 2.
[3] H. Woods, *Q. J. G. S.* vol. LII. (1896), p. 68, *ib.* vol. LIII. (1897), p. 377.

PISCES.

> *Lamna appendiculata* Ag.

MOLLUSCA.

Cephalopoda.

> *Ammonites peramplus* Mant.
> *Crioceras ellipticum* Mant.
> *Heteroceras reussianum* D'Orb.
> *Nautilus sublævigatus* D'Orb.
> *Scaphites Geinitzi* D'Orb.

Gasteropoda.

> *Avellana* cf. *Humboldti* Müll.
> *Cerithium Saundersi* Woods.
> *Emarginula sanctæ-catharinæ* Passy
> *Natica vulgaris* Reuss
> *Pleurotomaria perspectiva* Mant.
> *Trochus Schlüteri* Woods, D'Orb.
> *Turbo gemmatus* Sow.

Scaphopoda.

> *Dentalium turoniense* Woods

Lamellibranchiata.

> *Inoceramus Brongniarti* Sow. ?
> *Lima Hoperi* Mant.
> *Modiola Cottae* Roem.
>   „   *quadrata* Sow.
> *Ostrea semiplana* Sow. ?
> *Spondylus spinosus* Sow.

BRACHIOPODA.

> *Rhynchonella limbata* Schloth.
>   „   *plicatilis* Sow.
>   „   *reedensis* Ether.
> *Terebratula carnea* Sow.
>   „   *semiglobosa* Sow.

CRUSTACEA.

> *Pollicipes* sp.

ECHINODERMATA.

> *Cardiaster ananchytis* Leske
> *Cidaris* sp. (spine)
> *Cyphosoma radiatum* Sorig.
> *Echinocorys (Ananchytes) ovatus* Leske
> *Holaster planus* Mant.
> *Micraster breviporus* Ag.
> „   *cor-anguinum* Leske var.
> „   *cor-bovis* Forbes
> *Pentacrinus Agassizi* Von Hag.

ACTINOZOA.

> *Parasmilia centralis* Mant. ?

PORIFERA.

> *Camerospongia campanulata* Smith
> „   *subrotundata* Mant.
> *Coscinopora globularis* D'Orb.
> *Ventriculites alcyonoides* Mant.
> „   *impressus* Smith
> „   *mammillaris* Smith
> „   *radiatus* Mant.

Economics of the Middle Chalk.

The chalk is burnt for lime and spread on the clay soil to render it more open as well as to furnish constituents in which it is deficient. It is also used for the same purpose unburnt, and its action then is slower but more permanent.

Agriculturally the ground occupied by the Middle Chalk in this area is largely arable. But here and there, as near Royston, the open downs covered with short springy turf are still untouched by the plough.

THE UPPER CHALK.

General remarks. The Upper Chalk covers only a small portion of this district, and its outcrop is much obscured by Boulder Clay. Only the lowest zone is exposed in our area.

The change from Middle to Upper Chalk is sharp, and the abrupt line of demarcation appears to indicate that a new set of conditions came in, perhaps prefaced by a slight erosion of the underlying beds. The Upper Chalk sea must have been more charged with silica or more populated with siliceous organisms, for in it the bands of flint nodules and layers of thin tabular flints are very numerous and conspicuous, whereas they are altogether absent in the Lower Chalk. The origin of these flint nodules in regular layers is considered by some geologists[1] to have been brought about contemporaneously with the deposition of the calcareous mud, and somewhat in the following manner:—Wherever the lowest layers of water became supersaturated with silica, the decaying organisms strewn on the sea-floor at that period caused a precipitation of the silica, each organism forming a centre of precipitation and attraction. As the calcareous mud charged with silica in solution continued to be laid down the process went on, and the silica was abstracted from the overlying and underlying mud for a definite thickness determined by the strength of the force of attraction. When that limit of thickness was exceeded

---

[1] *Mem. Geol. Surv.*, Explan. Quart. Sheet 51 S.W., p. 70.

by the constant deposition of mud above, the segregation of silica round these centres on the now buried sea floor would cease; and the process would begin again at the level of the new and higher sea-floor on which organic matter was decaying. Thus a series of layers of flint nodules would be formed, each marking successive levels of the floor of the chalk sea.

The tabular flints which occur in horizontal layers may have been formed in somewhat the same way, but when they fill fissures, joints and fault-planes are certainly in most cases of later origin, and may have arisen by a process of lateral secretion after the partial or complete consolidation of the chalk.

There is however a widely-accepted theory that the bands of nodular flints owe their origin to a kind of concretionary action operating on the silica disseminated through the mass of the chalk subsequent to the formation of the latter[1].

The Upper Chalk is about 450 feet thick.

[1] For discussion of the question see Wallich, *Q. J. G. S.*, vol. xxxvi. (1880), p. 68. Sollas, *Ann. Mag. Nat. Hist.* 5th ser. vi. p. 437. Julien, *Proc. Amer. Assoc.* vol. xxviii. p. 359. Murray and Irvine, *Proc. Roy. Soc. Edin.* xviii. (1891), p. 229.

(1) ZONE OF *MICRASTER COR-TESTUDINARIUM*[1].

Characters. The chalk of this zone is a soft earthy limestone with occasional bands of harder rock, marly layers, and seams of flint, either tabular or in nodules. The beds are almost horizontal.

Occurrence and exposures. The outcrop of this zone lies immediately above that of the Chalk Rock just described, but owing to the great mass of Boulder Clay on the edge of the escarpment only a narrow belt is exposed, and that in the centre of the eastern part of the area. Near Royston, at Tharfield, Little Chishall, Reed, Barkway and Heydon there are pits in which sections are seen. The pits at the lime-kilns N.W. of Balsham abound in fossils enclosed in flints. Near West Wratting, Westley Water-less, and Stetchworth are more pits in which these beds are exposed.

Palæontology. The following fossils have been found in this zone in Cambridgeshire:—

MOLLUSCA.

Lamellibranchiata.

Inoceramus Cuvieri Sow. ?
Lima Hoperi Mant.
Pinna decussata Goldf.
Spondylus spinosus Sow.

---

[1] This zone was termed in the Geological Survey Memoir of the Cambridge area (Quart. Sheet 51, S.W. pp. 69, 136) the Zone of *Micraster cor-bovis* because the type species had not been well determined in this district. But in the Memoir on the Geology of London (1889) the zone fossil *M. cor-testudinarium* which is used to denote this horizon on the Continent is adopted by preference.

BRACHIOPODA.

Rhynchonella plicatilis Sow.
Terebratula carnea Sow.
    „   *semiglobosa* Sow.
Terebratulina gracilis Schloth.
    „   *striata* Wahl.

ECHINODERMATA.

Cidaris sceptrifera Mant.
Cyphosoma radiatum Sorig.
Echinocorys (Ananchytes) ovatus Leske
Epiaster gibbus Lam.
Micraster cor-anguinum Leske?
    „   *cor-bovis* Forbes
    „   *cor-testudinarium* Goldf. (var. *brevis*)

ACTINOZOA.

Parasmilia centralis Mant.

PORIFERA.

Alecto sp.
Coscinopora globularis D'Orb
Ventriculites sp.

THE PLEISTOCENE DEPOSITS.

There is a gap in the geological sequence over this area between the Zone of *Micraster cor-testudinarium* of the Upper Chalk and the Glacial Drift. The uppermost beds of the Upper Chalk have all been removed by denudation; and all the Tertiary beds are now absent, though some of them may possibly have once spread over the district. The late Pliocene beds which usher in the Glacial Period in the adjoining county of Norfolk and which are so well seen on the coast are completely wanting; and even the earlier glacial deposits are not represented, save possibly by some unimportant local patches of gravel at or near the base of the Boulder Clay.

The Pleistocene deposits may be conveniently treated under the following four heads :—

(1)   The Glacial Drift including the Boulder Clay and the High Level Gravels.

(2)   The River Gravels.

(3)   The Fenland deposits.

(4)   Recent Alluvium, warp and trail, etc.

THE GLACIAL DRIFT.

THE BOULDER CLAY.

Characters and thickness. In this area the Boulder Clay, which belongs to the Great Chalky Boulder Clay of East Anglia, consists mainly of a dark grey or bluish clay weathering to a drab or brownish colour to the depth of several feet. Locally its colour and contents vary according to the rocks on which it rests or which occur in the vicinity. Thus where it rests on the Chalk it consists mainly of ground-up chalk, and the matrix is therefore very calcareous; but where it lies on the Gault or other clays it is composed of much worked-up argillaceous material and is frequently scarcely distinguishable at first sight from the parent bed. The Boulder Clay however hardly ever shows any stratification.

Fragments of hard chalk are usually very abundant, and may be found in it almost everywhere after a little search; they are more or less rounded or sub-angular in shape and often show flattened and striated surfaces; they range in size from a mere grain up to a large pebble. Occasionally there are included masses of Chalk of much greater size, and at Ely an enormous block of Chalk and other Cretaceous strata lies imbedded in the Boulder Clay which occupies a small valley scooped out of Kimeridge Clay (see p. 176).

In addition to these lumps and masses of Chalk, boulders of other rocks foreign to the area frequently occur, and are scattered promiscuously through the Clay. Amongst these boulders there are recognised blocks of Carboniferous sandstone, Mountain Limestone, Oolitic limestones, Red Chalk (probably from Hunstanton or the Yorkshire coast), phosphatic nodules from the Cambridge Greensand, carstone, Chalk flints, slates, hornstones, and metamorphic and igneous rocks, such as gneiss, mica-schist, quartzite, porphyrite, basalt, granite, diorite, etc. These blocks are of various sizes; two found at Long Stanton measured 3 × 3 × 2 ft. They sometimes show the smoothed, polished, or grooved faces characteristic of glaciated rocks. Many are smooth, well-rolled pebbles, resembling beach-shingle. Fossils derived from the Lias, Oxford Clay, Kimeridge Clay, Cambridge Greensand, Chalk, etc. are not uncommon.

Interbedded lenticular patches of gravel, sand, or loam do not often occur in the Boulder Clay of our area. There is one instance in the railway-cutting west of Newnham Hall between Ashdon and Bartlow where a bed of loam is found intercalated in the clay; and again near the top of Barrington Hill near Linton where a thin bed of laminated loam similarly occurs.

In the neighbourhood of Linton small deposits of gravel or sand are found separating the Clay from the Chalk, but they are rare and only local. The base of the Clay is usually sharply defined from the subjacent rock, except where that is also clay and has been much worked up. The chalk-floor beneath the Clay does not show any striated or polished surface owing to its soft nature and the removal of the original surface by subterranean denudation. The basement layer of the Clay is sometimes of a sandy nature, but where the Clay rests on the Chalk there is sometimes a thin band of hard calcareous rock. It has been suggested[1] that this hard band is due to the

[1] *Mem. Geol. Surv.* Explan. Sheet 47, p. 60.

carbonated water, which after percolating through the Chalk, threw down its excess of bicarbonate of lime where its free passage sideways or upwards was arrested by the impervious layer of Boulder Clay. This explanation is probably correct, for the hardened band is mostly present where the Clay occupies an old hollow in the surface of the Chalk. The band may be seen in a chalk-pit, 1 mile south of Bartlow Station, and in the railway cutting west of Dullingham Station.

The thickness of the Boulder Clay necessarily varies much, for it has not only suffered much denudation since its formation but it was originally deposited on an uneven surface, so that where it filled up old hollows or troughs it is of considerable thickness, as at Impington, where it has recently been proved to be 60 ft. thick[1]. Unequal earth-movements of a later date have probably operated to depress these troughs below the sea level. On the hill tops and level ground the Boulder Clay is much less thick, partly owing to erosion, but minor variations in its original thickness possibly always existed even on comparatively level surfaces owing to irregularity of de-position or accumulation.

The amount of glacial deposits over this area must have been enormous. Prof. Bonney[2] states that "at Old North Road Station on the Cambridge and Bedford Rail-way as shewn in the cutting and well it [the Boulder Clay] was 160 ft." thick. Near Caxton and Longstowe it is proved by wells to be from 160—180 ft.; while in the village of Caxton half a mile to the north of the deep

---

[1] In the upper part of the valley of the Cam in Essex the bottom of the Drift in one place was not reached at a depth of 340 ft. or nearly 140 ft. below sea-level.

[2] Bonney. *Camb. Geol.* p. 49.

sinking it is only 14 feet[1]. At other places it has diminished to 2 or 3 ft. in thickness. Thus it appears that on the pre-glacial surface of the country there existed numerous deep channels and troughs of which no trace is now seen at the surface, and the presence of which is only detected by borings or deep cuttings.

Occurrence. The Great Chalky Boulder Clay covers an area of 3000 square miles and reaches an altitude of 500 feet above sea-level. It stretches from the East coast westwards to Abbots Bromley in Staffordshire. In the Trent valley it is underlaid by beds of gravel, sand, and loam which in places are interstratified with the lowest portion of the Clay and sometimes contain fragments of marine shells. West of the Trent basin the Chalky Boulder Clay dies out and we meet with a new Boulder Clay composed of rock detritus and boulders from the north-west, which according to J. Geikie[2], was the ground-moraine of a lobe of the great Irish Sea glacier with feeders from the Lake District, Wales, and the Pennine Chain[3].

Of the Cambridgeshire area the Boulder Clay covers

[1] Seeley. *Q. J. G. S.* vol. XXII. (1866) p. 471.

[2] J. Geikie. *The Great Ice Age*, 3rd Ed. pp. 374—375.

[3] Some geologists (see *Geol. Mag.* Dec. IV. vol. III. 1896, p. 449 and references) contend that there is in reality only one Boulder Clay in Eastern England and that the various clays which have received different names and are supposed to be of different ages are merely its local developments. Thus they hold that the Chalky Boulder Clay graduates horizontally into the Stony Loam or Lower Boulder Clay of Norfolk and into the Purple Clays of Lincolnshire and Yorkshire. The separation of the several clays and their order of superposition is a matter of considerable difficulty, but the majority of geologists intimately acquainted with the drift deposits of Eastern England maintain that these clays are not to be regarded as contemporaneous or local varieties of one extensive sheet.

all the western side, capping the high ground on the
north of the Rhee valley and spreading in a sheet over
the upland region between the valleys of the Ouse and
Cam. It sends a long narrow promontory eastwards to
Haslingfield; and further north another much larger and
broader mass projects in the same direction with an
irregular boundary running roughly from Kingston to
Coton and Madingley, thence in a north-westerly
direction to Dry Drayton and Lolworth, and after
throwing out a long cape towards Conington passing
through Elsworth and Papworth into Huntingdonshire.
North of St Ives the Boulder Clay caps all the high
ground and extends down to the margin of the fens.

In the Fenland itself there are several 'islands' more
or less composed of Boulder Clay and rising above the
general level of the plain. There is the large connected
group of patches between Sutton, Wentworth, Ely, and
Chettisham, where the Boulder Clay caps the low ridges
and hills of Lower Greensand, Gault, and Kimeridge Clay.
At Ely its base is about 60 feet above ordnance datum.
Further north there are many islands, such as those
of Manea, Pye Moor, Stonea, Butchers' Hill, Littleport,
and March, which consist almost exclusively of Boulder
Clay. It is possible that "all these island masses may
be connected underground and form one long channel."

In the eastern part of our area we only meet with
the western fringe of the great East Anglian sheet of
Boulder Clay. This sheet lies on the top of the Chalk
escarpment with a very irregular boundary and extends
thence south-eastwards to cover large parts of Essex
and Herts. The general trend of the boundary is from
Tharfield eastwards to Saffron Walden, thence northwards past Great and Little Chesterford to Linton, and

on the further side of the Lin Valley north-eastwards, past Balsham, West Wratting, Dullingham, and Cheveley into Suffolk.

Contour of surface on which the Boulder Clay rests. A review of the various heights at which the Boulder Clay rests on the underlying formations reveals the fact that the Cambridge valley was excavated in preglacial times[1].

At Tharfield in the south the base of the Boulder Clay lying on the summit of the escarpment is 530 feet above the sea. A gradual and fairly regular downward slope to the north-east following the line of the Chalk escarpment is found to exist, for the Boulder Clay lies at a level of about 350 feet at Balsham and at 240 feet at Dullingham. The gradient is in this direction about 10 feet in a mile, but in a transverse direction at right-angles to the trend of the escarpment the base of the Clay descends much more rapidly, and by connecting together the various points on which Boulder Clay rests along a transverse line drawn through Tharfield, Cambridge, Long Stanton and the high ground north of St Ives this fact is conspicuously brought out. The line connecting these points is seen to be a uniclinal curve, and this must correspond with the old surface on which the Boulder Clay was laid down[2]. The steepest slope of this surface (1 in 57) is found to lie somewhat in front of the present escarpment, while beyond this point the slope rapidly decreases to 1 in 264, and then 1 in 330, but finally becomes 1 in 300. Beyond Long Stanton and

---

[1] *Mem. Geol. Surv.* Explan. Quart. Sheet 51 S.W. p. 75. Penning, *Q. J. G.S.* xxxii. (1876) p. 196.

[2] *Mem. Geol. Surv.* Explan. Quart. Sheet 51 S.W. pp. 75—78.

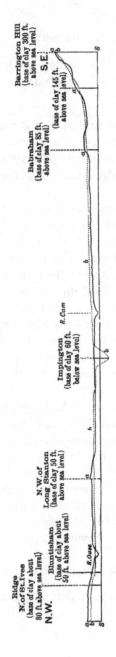

FIG. 10. DIAGRAM-SECTION SHOWING CONTOUR OF LAND IN GLACIAL TIMES.

Vertical scale about 800 ft. to 1 inch.

Length of section about 24 miles.

SS   Present sea level.

aa   Present contour of land.

bb   Contour of land when Boulder Clay was deposited.

towards Bluntisham the base of the clay rises with a
gradient of about 1 in 88. (Fig. 10.)

Thus we see that the Cambridge valley must have
existed very much in its present form before the Boulder
Clay was deposited in it. The upper part of the Cam
valley has also been described as pre-glacial[1], and recently
Mr Whitaker[2] has brought forward evidence pointing to
the existence of a deep pre-glacial channel in the Chalk
extending past ˙Audley End and Saffron Walden to
Quendon, in which the Boulder Clay attains a thickness
of 218 feet. There is also evidence that some of the
tributary valleys are pre-glacial. Thus the present valley
of the Lin (or Bourn) between Linton and Babraham
nearly occupies the line of a valley partially filled with
Boulder Clay. This is indicated by the base of the Clay
being 350 feet above sea-level at Balsham, 300 feet at
Barrington Hill, and 120 feet at Abington, while to the
south it rises again to 400 feet. Near Chesterford there
is similar evidence of an old valley.

Mode of formation of the Chalky Boulder Clay.

Any theory which would explain the origin and
method of formation of this deposit must account for
the following facts:—

(1) its distribution and relation to the present contour
of the country, capping the high ground, descending into
the valleys, filling up old river-courses, and spreading out
over the plains.

(2) its composition, viz. its dependence for its lithological characters and contents on the underlying rock;
its local variation arising from this cause; the presence

[1] *Mem. Geol. Surv. Explan. Sheet* 47, pp. 36--38. f. 12.
[2] *Q. J. G. S.* vol. XLVI. (1890), p. 333.

of an immense quantity of fragments of Chalk of all
sizes, from mere dust up to boulders many feet in length ;
the angular or subangular shape of these fragments with
their occasionally scratched, facetted, flattened and scored
surfaces ; the presence of other boulders and pebbles of
non-local or foreign origin ; the presence of well-rolled
pebbles resembling beach shingle.

(3) the passage of material derived from underlying
rocks in the north-east over the outcrop of other underlying rocks to the south-west, as is shown by the overlap
of the detritus from the Chalk on to the Kimeridge Clay.

The question of the formation of this Clay should not
be considered with reference to the Cambridgeshire district
only, but after a general review of the behaviour and characters of the deposit over the whole area of its extension. Two
facts, however, of local importance must be borne in mind ;
firstly, that the so-called Lower Glacial beds of the counties
to the east are wanting in our Cambridgeshire area ; and
secondly, that there is likewise an almost complete absence
of the so-called Middle Glacial beds which are so well
developed in the adjoining country to the south and east
of the Chalk range of hills against which they thin out or
extend only into a few of the deeper transverse valleys, as
possibly at Chesterford and Bartlow.

The existence of the Cambridge valley and of some
of its tributary valleys before the accumulation of the
Chalky Boulder Clay rests on the very strong evidence
given above, so that we must bear in mind that the
general relief of the country cannot have been very
different to that which we find to-day, though this does
not necessarily preclude the idea that as a whole it stood
at a lower or higher level.

The chief theories put forward to account for the

origin and distribution of the Chalky Boulder Clay fall into two groups, (i) the glacial, and (ii) the non-glacial. In the first group there are four theories :—(1) *the ice-foot theory;* (2) *the ice-berg theory;* (3) *the extra-morainic lake theory;* (4) *the land-ice or ice-sheet theory.* In the second group there is only one of importance and it may be called *the diluvial theory.*

All these theories call in earth-movements of greater or less magnitude and intensity. But while some require elevation of the land, others demand submergence, as will be pointed out.

It must be confessed that none of the theories are completely satisfactory or account for all the peculiarities of the deposit.

The *ice-foot theory*[1] is put prominently forward in the Geological Survey Memoir (*Quart. Sheet* 51, S.W.) of this district, and in works by Mr Jukes Browne. According to this theory the subsidence of the country which began in mid-glacial times was not sufficient to allow the sea to pass over the top of the Chalk escarpment to flood the Cambridge valley during the accumulation of the Middle Glacial sands and gravels, so that they were only piled up on the eastern or seaward side of the Chalk slope[2], while the supposed continuation of the Chalk ridge across the Wash during this period is believed to have prevented the deposits drifting into the mouth of the Cambridge valley. Into a few, however, of the deeper inlets (*e.g.* Chesterford and Bartlow) of the sinking coast-line the gravels may have been carried.

---

[1] A. Jukes Browne, *Building of the British Isles* (1888), p. 287 et seq. *Handbook of Historical Geology* (1886), p. 564.

[2] *Q. J. G. S.* vol. xxxiii. (1876), p. 191; vol. xli. (1885), p. 130. *Mem. Geol. Surv. Explan. Quart. Sheet* 51, S.W. p. 115.

On the return of glacial conditions which occurred during the progress of the subsidence an "ice-foot" formed all round the shore, such as travellers have described in the Arctic regions, where the first frost of the late summer covers the surface of the sea with a crust of ice which is carried upward by the rising tide and becomes glued to the rocks or cliff-face, thus growing in thickness with each successive tide. On a shelving and sinking shore covered with a thick layer of rotten surface-rock the ice-foot tears up great frozen masses of stones, soil, and solid rock which grate and grind over the slope, or, imbedded in the ice, are floated away with the bergs as the ice melts or gets broken up. This detritus is ultimately deposited on the sea-floor. The sediment carried by marine currents or washed down into the sea by land streams becomes mixed with this ice-borne material and forms irregular stratified patches of loam, gravel, or sand.

Landslips and avalanches from the cliffs materially help in the rapid and tumultuous accumulation of materials in the sea in Arctic regions, and the floe-ice and ice-bergs which are driven on the shore by currents and storms ground in the shallows or on sunken ridges, tearing up and pulverising the solid rock, as well as depositing their far-travelled burdens of débris. Any pre-existing beds of soft unconsolidated material would be worked up and incorporated into the freshly formed glacial deposits, except where they lay in protected positions or at considerable depths.

It is pointed out that the gradual submergence of the whole land-area would cause the various formations which successively formed the shore-line to be subjected to this destructive action of the ice-foot and to the grounding of the ice-floes. It is also thought that the sea over this submerged part of England was to some extent shut in to the south, so that there existed a sort of huge bay into which the ice-floes and bergs would be driven and packed together, and by that means the intensity of their grinding action be increased.

It is to such conditions and processes that some geologists would ascribe the formation of our Chalky Boulder Clay. The usual objections urged are : (1) there is no definite evidence of marine conditions; (2) it has never been indisputably shown that the coast-ice of the Arctic regions tends to form such an extensive deposit; (3) the distribution of the Clay is not what we should expect if we bear in mind the configuration of the ground on which the materials were laid down ; (4) the resemblance of the Clay to deposits elsewhere which are generally held to have been made by extensive glacier-ice would *primâ facie* incline us to ascribe its origin to some similar process as that by which they were formed[1].

The second theory is that which calls in the agency of *ice-bergs* floating over a depressed area and dropping their burdens on the sea-floor. In the first place this theory necessitates a submergence of the land to an extent greater than the height (500—600 ft.) of any of the hills which the Clay caps, unless we imagine differential depression. It is supposed that in the sea which spread by this submergence over Eastern England there floated huge masses of ice laden with débris which was deposited on the sea-floor as the bergs melted or grounded in the shallows. As these bergs drifted along they caught on the ridges and sunken reefs, tearing off great masses of

[1] S. V. Wood and F. W. Harmer. *Q. J. G. S.* vol. xxxiii. (1877), pp. 115—119 and vol. xxxviii. (1882), p. 667. J. Geikie, *Great Ice Age*, 3rd ed. 1894, pp. 342—352.

the harder solid rocks and ploughing up the softer beds. This local material was pounded and churned up beneath the water by the collision and stranding of the bergs, and when the ice-floes melted and glacial conditions passed away it was mixed with the débris either imbedded in or resting on the ice.

There are many serious objections to this theory. (1) The material brought by ice-bergs comes from distant gathering grounds and therefore the clay formed by such agents should show mainly foreign and non-local cha-racters, which is exactly the reverse of what we find in the Chalky Boulder Clay; (2) the almost continuous sheet of clay over hill and valley is quite unlike the local heaps of detritus which ice-bergs would throw down as they melted or toppled over on grounding; (3) the relation of the composition of the Boulder Clay to the underlying rocks is almost impossible to explain on this ice-berg hypothesis; (4) the almost complete absence of remains of marine organisms over large areas is another but not so serious a difficulty. The sea must have been fairly open, for if it had been permanently ice-bound, a sufficient number of ice-floes laden with detritus could not have drifted in a constant stream or in successive batches to have given rise to such a thick and widespread deposit as this Boulder Clay. (5) The want of stratification in the deposit is hard to explain, as there must have been some kind of sorting of the materials, at any rate by gravity, as they sank to the bottom[1]; (6) the invasion of the outcrop of one rock by the detritus of the adjacent

[1] Col. H. W. Feilden, (*Q. J. G. S.* vol. LII. (1896), p. 52) who holds that the Boulder Clays of Kolguer Island are of glacio-marine origin, states that they show no definite stratification; and he compares them to our Chalky Boulder Clay.

rock in regular order and succession from north-east to south-west is another feature difficult to account for satis-factorily on this hypothesis; (7) the configuration of the country would have kept out the largest bergs from the area where the Clay is now thickest, and would thus have cut off part of the supply of material.

The only important point which this theory explains satisfactorily is the miscellaneous assortment of foreign or non-local boulders which we find mixed with the local materials. Ice-bergs are borne along principally by marine currents, and a small berg of small draught may be drift-ing in the surface-current in one direction while a giant berg with deep draught may float along in an opposite direction under the influence of a more powerful under-current. Thus the rock-fragments which they bring along and ultimately deposit in the same spot may come from most diverse localities. But the transportation and inter-mixture of the foreign erratics can also be explained by the land-ice theory in nearly as satisfactory a way (see sequel).

The *extra-morainic-lake theory* was put forward by Prof. Carvill Lewis[1]. He held that a great lake existed in front of the huge ice-sheet which covered the north and north-west of England and which was confluent with the still larger ice-sheet of the North Sea. This ice-sheet, 300 ft. thick at its edge, dammed back all the rivers flowing out to sea and caused the formation of a huge inland freshwater lake—an "extra-morainic lake"—extending from York to Harwich. Numerous rivers flowed into it on the south and west, and ice-bergs detached from the edge of the ice-sheet floated across its surface and dropped their heavy loads of débris on its floor. Vast quantities

[1] *The Glacial Geology of Great Britain*, etc. p. 55 et seq.

of mud were contributed by the subglacial rivers and by the glaciers themselves. The ice as it formed round the shores of the Lincolnshire Wolds which stood up as islands above the water carried out great cakes of Chalk and deposited them in the lake. England, according to Prof. Lewis, then stood 100 ft. lower than it does to-day, so that with the ice 300 ft. thick at the edge the lake must have reached a depth of nearly 400 ft. When it had attained almost this depth it may have burst its bounds, whereupon the waters would have rushed as a genuine *débâcle* over the country to the south-west, carrying enormous quantities of gravel and sand over the land. As the ice melted and retreated northwards and the level of the lake gradually sank, 'the deepest and densest mud would be deposited nearest the glacier in the deepest portion of the lake; the deposits on the other hand at the inland borders of the lake would be thin and stony, and at the entrance of the rivers would be replaced by fluviatile sands and gravels.'

The chief objections to this theory are (1) that it fails to explain the peculiar local characters and distribution of the Chalky Boulder Clay; (2) that great warpings of the earth's crust and elevation of local ridges have to be assumed in order to provide dams to retain the waters of this lake at the requisite height; and (3) that these supposed lacustrine and subaqueous glacial deposits show no strong peculiar features by which they can be distinguished from those which are held by the same author to have been formed by land-ice.

Mr Searles Wood[1] held a modification of this theory combined with that of the land-ice, for he maintained that this Clay was extruded from the foot of an ice-sheet

[1] *Q. J. G. S.* vol. xxvi. (1870), p. 100.

and deposited under water. The general absence of stratification in the drift is however against every theory of a subaqueous origin, particularly in the light of Mr M. J. C. Russell's[1] recent investigations on the morainic materials deposited in the sea at the foot of the Malaspina Glacier in Alaska.

We now come to the theory which is most widely accepted—the theory of *land-ice*. Though it fails to offer a completely satisfactory explanation of all points, yet it presents perhaps fewer difficulties than any of the foregoing. The evidence of the action of land-ice as glaciers or ice-sheets over the northern and western parts of Great Britain, the Continent, and North America is very strong, and owing to the many points of similarity in character and mode of occurrence it seems impossible to escape the conclusion that we have here also to deal with a deposit formed by the same agents[2].

The great glaciers which descended from the mountains of the Lake District and rode across Yorkshire were joined by prolongations of the Scotch glaciers, and this whole mass of British ice being crushed against the Yorkshire coast, south of Bridlington, by the pressure of the more powerful Scandinavian ice-sheet was forced to take a south-westerly course. This ice, together with the front of the Scandinavian sheet, flowed over the whole of East Anglia with a great ground-moraine beneath it, composed principally of the rocks on which it rested,

[1] 13th *Ann. Rep. U. S. Geol. Surv.* (1891—1892), pt. II. G. F. Wright and Warren Upham, *Greenland Icefields* (London 1896), chap. XI.

[2] For a summary of the question see Dr Wright's *Man and the Glacial Period* (*Internat. Scientif. Ser.*), p. 158 et seq., and for an excellent criticism of the theories see Prof. Bonney's *Ice-Work* (*Internat. Scientif. Series*), 1896, p. 163. Also J. Geikie, *The Great Ice-Age*, 3rd ed. 1894, chapter XXV.

for the long exposure of the land surface of this region to atmospheric weathering in late Tertiary times had prepared a thick layer of rotten rock and subsoil. Moreover each outcrop of solid rock supplied materials not only to the moraine which rested directly upon it but also to that on the next outcrops to the south-west, the detritus being pushed along in that direction. The loose and unconsolidated deposits of Tertiary age, which formerly had a much wider extension, must also have furnished many of the rounded beach-pebbles of the Clay.

Since the main movement of the ice-front was against the drainage of the country there must frequently have been extra-morainic lakes formed, over which icebergs floated, dropping their loads in irregular patches. In these lakes deposits of sand and loam accumulated in stratified beds. Local floods, due to the bursting and overflow of these temporary lakes when they rose above the level of their shores or burst their barriers, caused irregularities of deposition and distribution and even local erosion, and led occasionally to the intercalation of more or less stratified beds in the midst of the ground-moraine.

The ice carried on its surface or imbedded in its mass boulders of far-travelled rocks which it had picked up in its long journey from the north or east, and these were incorporated with the local materials. But their transportation was in many cases a very complicated process, and possibly ice-bergs in early glacial times helped to scatter them over the sea-floor, to be subsequently picked up or pushed along by the advancing glaciers. Some may have been derived from the Cambridge Greensand, in which foreign boulders are found (see p. 103). The solid rock here and there, owing to some local configuration of the ground, or to currents in the ice, or to other causes, appears to have been not only denuded of its loose superficial deposits by the scour of the ice but also pounded up into fragments ; and in some cases huge blocks of strata seem to have been sheared off and carried along some distance from their home before being dropped in the soft clay. Examples are numerous in the Contorted Drift of Cromer, into which the Chalky Boulder Clay appears to pass laterally[1].

The absence of Middle Glacial beds over Cambridge-shire and their presence on the south-eastern side of the Chalk hills is explained by Messrs Wood and Harmer[2] by the supposition that during the period of their accumu-lation the land-ice rested over the Cambridgeshire area, and the streams which issued from the base of the ice washed out and distributed the gravels over the sea-floor, so that they are now found far beyond the limits of the Upper Glacial (the Chalky Boulder Clay) in S.E. Suffolk and northern Norfolk.

When the land-ice began to retreat northwards from this region the ground-moraine was left behind as a continuous sheet of clay, while the streams issuing from the front of the retreating glaciers must have been powerful agents of denudation.

The occasional preservation of gravel beds below the Boulder Clay, such as those at Chesterford and Bartlow, may perhaps testify to their former wider extension and to the early destructive powers of the ice-sheet. Or we may regard them as representing the deposits of small local extra-morainic lakes which existed before the Boulder Clay was formed.

[1] *Mem. Geol. Surv., Geol. of Cromer*, p. 111 *et seq.*
[2] *Q. J. G. S.* vol. xxxiii. (1877), pp. 115—119, and vol. xxxvii. (1882), p. 667.

Mr Searles Wood has been inclined to invoke the action of a local glacier or group of local glaciers rather than the Scandinavian ice. But it is difficult to see how a glacier of sufficient size could have gathered on the low hills of eastern England, and also how we are to account for the presence of the foreign and non-local erratics.

One of the chief difficulties adduced against the acceptance of the land-ice theory is the great elevation at which much of the Boulder Clay rests. In order that it might be possible for the edge of the Scandinavian ice-sheet to mount the slopes of the Chalk escarpment, the "head" of ice and the pressure from behind must have been enormous. The presence, moreover, of the great channel 400 fathoms deep off the Norwegian coast seems to offer an impassable obstacle to the passage of the ice-sheet in a westerly direction. Some glacialists get over this difficulty by declaring that the ice could and did cross it; others assert that the channel was not formed till post-glacial times; but this is a highly improbable assumption.

The above-mentioned difficulty of the height of the Boulder Clay above sea-level (*i.e.* 530 feet at Tharfield, 300 feet in Suffolk, and about 175 feet close to London) is a very serious one, for the source of the ice was some 200—300 miles distant, and for it to have ignored the configuration of the ground so close to its margin seems almost incredible. But there is a possibility that East Anglia stood at a lower level with reference to the North Sea at that time, and the submergence of eastern England as a whole at this period may have been sufficient to prevent the then existing hills being any obstacle to the onward course of the ice. On this hypothesis the re-elevation of the district seems to have accompanied or immediately followed the departure of the ice.

Another difficulty is the mixture of rocks from all parts of England and the presence of the foreign and non-local boulders in the clay. Some of these were probably derived from earlier beds existing over the area which the ice traversed and from the surface of which it ploughed up and scoured off most of the loose or weathered materials. The intercrossing of erratics[1] and their upward or downward movements in the ice[2] are explained by the existence of currents in the ice itself, and we may thus account for, at any rate to some extent, the strange intermixture of rocks.

In many places in East Anglia the strata under-lying the Chalky Boulder Clay are much disturbed and contorted, and this is brought forward as evidence of the passage over them of a heavy mass of ice. The crumplings in the Chalk are pointed out as the direct results of the pressure of the ice.

There has always been much dispute about the conditions of formation or even existence of a ground-moraine comparable in thickness and characters to our Boulder Clay. If we admit at all the existence of a ground-moraine capable of giving rise to such a deposit, we may reasonably imagine it would be spread over the comparatively level country where the ice had lost much of its abrading power in consequence of the lower gradient[3]. Recent investigations of Arctic ice-sheets do not contradict this view, but they also

---

[1] Prof. James Geikie, *Fragments of Earth Lore*, p. 194.

[2] Sollas, *Q. J. G. S.* vol. LI. (1895), p. 361. Prof. Bonney, *Ice Work*, p. 184.

[3] See S. V. Wood, *Q. J. G. S.* vol. XXXIII. (1877), p. 115.

emphasise the importance of the "englacial" material[1]
(*i.e.* the material imbedded in the general mass of
the ice) and assert that it equals or even exceeds the
amount of sub-glacial débris. On the melting and
retreat of the ice-sheet this included material would
contribute largely to the morainic material left behind,
and may account for many of the far-travelled boulders.
By the retreat of the glaciers in Alaska and Greenland
a sheet of material comparable to the Boulder Clay is
left spread over the ground[2].

The last theory to be mentioned is that which may
be called *the diluvial theory.* It has been recently
strongly and ably advocated by Sir Henry Howorth[3]
as alone furnishing a satisfactory explanation of the
formation of the Chalky Boulder Clay. According to
this writer the Clay is the result of great floods combined
with sudden and violent earth movements. "The denudation of the Fen Country which produced the great
mass of the Chalky Clay with most of its boulders was
coincident with and caused by the bending and folding
of the Chalk of eastern England which took place after
the deposition of the Crag beds, and during the period
of folding a great depression was formed around the
Wash into which the water rushed from the north
carrying débris and mixing it with clay; this rushing
into what was virtually a cul-de-sac dispersed and
scattered its load in all directions."

---

[1] Warren Upham, *American Geologist*, vol. xii. 1893, p. 36.
[2] See J. Russell, 13th *Ann. Rep. U. S. Geol. Surv. loc. cit.* and Wright
and Upham's *Greenland Icefields.*
[3] *The Glacial Nightmare and the Flood* (1893), chapters xvii. and xviii.
Q. J. G. S. vol. li. (1895), p. 504.   *Geol. Mag.* Dec. iv. Vol. iii. (1896),
pp. 58, 298, 449, and 553 ; *ibid.* vol. iv. (1897), pp. 123, 153, and 213.

The "violent disintegration and dislocation" of the Chalk was "rapid and cataclysmic" and was "the result of some great strain suddenly applied." The evidence offered in favour of this is the abundance of shattered chalk fragments in the Clay, and the occurrence of local disturbances and crumplings of the beds of Chalk in the area and as far to the east as Denmark. The huge isolated masses of Chalk etc. now widely separated from the parent rock and often imbedded in the Clay are considered by Sir H. Howorth to indicate dislocation of once continuous beds and not transport by ice.

This diluvial theory finds few supporters at the present time. Apart from the disinclination of modern geologists to invoke catastrophic action to account for widespread deposits (though Professor Prestwich[1] did so recently to explain the origin of the "Head" of the South of England and of similar deposits elsewhere), there are definite grave objections brought forward against it. It is urged that it fails to explain the relation of the contents of the Clay to the underlying rocks; the presence of the foreign boulders; the similarity of the deposit in many ways to undoubted glacial deposits elsewhere; the action of floods spreading the material over hill and vale with such remarkable independence of the configuration of the ground; the extremely local character of the "violent" earth movements postulated; the source of the water to cause the flood and the place where the currents originated; and the absence of sorting in the materials, since currents sufficiently powerful to have swept down enormous boulders

---

[1] *Q. J. G. S.* Vol. xlviii. (1892), p. 263. *Proc. Victoria Institute*, vol. xxvii. (1893), p. 263.

would have carried the fine gravel and sand to much greater distances. There is also the physical improbability or even impossibility[1] of such rapid earth movements as those postulated. Sir H. Howorth's acute criticisms of other theories are however deserving of careful considera-tion, for there are many unsolved problems and perplexing questions with which we have to reckon whatever theory we adopt. The land-ice theory under various forms finds the largest measure of support amongst modern geologists[2].

Roslyn Pit, Ely.

The extraordinary and anomalous position of the strata in this pit was formerly explained by a complex set of faults[3], but Prof. Bonney[4], following up the idea first put forward by the Rev. O. Fisher[5], completely showed that the fault theory was untenable, and that we must regard the mass of Cretaceous beds as a huge boulder transported by ice and dropped in a valley of Boulder Clay. This view has been generally adopted[6].

[1] Warren Upham, "British Drift Theories," *American Geologist*, vol. xiii. (1894), p. 275.

[2] See J. Geikie's *Great Ice Age*, 3rd ed. 1894. Dr Wright's *Man and the Glacial Period*, 1892. H. B. Woodward's *Geology of England and Wales*, etc.

[3] Sedgwick, *Rep. Brit. Assoc.* 1845, p. 42. Seeley, *Geol. Mag.* vol. i. 1864, p. 150; vol. ii. 1865, pp. 262, 529; vol. v. 1868, p. 347.

[4] Bonney, *Camb. Geol.* Appendix ii. p. 69. *Geol. Mag.* vol. ix. 1872, p. 403.

[5] Fisher, *Proc. Camb. Philosoph. Soc.* 1867, vol. ii. pt. v. p. 51. *Geol. Mag.* vol. v. 1868, 407.

[6] Skertchly, *Geol. of Fenland Mem. Geol. Surv.* p. 236. Whitaker and others, *Mem. Geol. Surv. Explan. Quart. Sheet* 51, N.E. 1891, p. 63.

For details of the refutation of the fault theory Prof. Bonney's account may be consulted. He showed that after allowing every circumstance favourable to simplicity there would be requisite two reversed faults with an ordinary downthrow fault outside each—a phenomenon of an unprecedented and most improbable nature in this district. The clay-pits, known collectively as Roslyn Hill Pit, or Roslyn or Roswell Hole, are situated about a mile E.N.E. of Ely Cathedral by the side of the railway line to Huntingdon. The great excavation stretches in a south-easterly direction from the base of the outlier of Lower Greensand down the slope of the underlying Kimeridge Clay. It is usual to treat of the pits in three portions for the sake of convenience. The first portion, known as the *Great Pit*, lies to the north-west of the railway and is the one generally visited because the sections are most distinct and accessible. The *Middle Pit* lies east of the railway between it and a by-road and is much obscured by vegetation. The *Lower Pit* is the most eastern, and being at or near the level of the fen is generally filled with water.

In the *Great Pit* (see Bonney, *Camb. Geol.* p. 74, fig. V.) the sections exposed in the sides are those in which the structure can best be studied, for the floor is occupied by ponds and partly overgrown with rushes. On the north-eastern side the Kimeridge Clay (p. 41) is seen in place in almost horizontal beds containing bands of large septaria and capped by thin gravel which lies immediately beneath the soil. The Kimeridge Clay is also seen at the northern and southern ends of the western side. At the southern end it is bent downwards towards the great erratic at an angle of about 20°. At the north-western corner of the pit the junction between the Kimeridge Clay and

Boulder Clay is obscured. The Boulder Clay and erratic lie lengthwise in a small valley ploughed out of the Kimeridge Clay and extending from the house called 'Little London' on the west to the River Ouse on the east. The cross section of this valley with its banks of Kimeridge Clay is now exposed on the western side of the Great Pit. The structure and relations of the erratic are best understood from a study of the observations which have been made upon the west and east faces.

In · the extreme south-west angle of this pit the Kimeridge Clay was seen to dip beneath the Boulder Clay at an angle of 20° to the north. A mass of shattered Kimeridge Clay separated the undisturbed Kimeridge Clay from the erratic, and the junction of the shattered and undisturbed Kimeridge Clay was a slickensided surface inclined at an angle of 60°. The other side of the shattered clay was bounded by a similar slickensided surface inclined at an angle of 30°. These slickensided surfaces have been mistaken for fault planes.

On this surface rested a more or less shattered mass of Lower Greensand which appeared to have slipped down the clay-slope. It was partly mixed with Boulder Clay, and the dip was determined to be 30° in an easterly direction. This mass is no longer exposed, but the overlying mass of Boulder Clay contains fragments of its fine ferruginous conglomerate, and they also lie scattered on the talus slope. Resting on this clay is the erratic, the basal beds of Gault and the rest consisting of the successive beds of the Chalk in their usual sequence, and but little disturbed. In fact the beds of which it is composed are almost horizontal, and thus do not agree in dip or strike with the above mentioned Lower Greensand and are not conformable to it. This narrower end of the great erratic is no longer plainly visible on the western face, but is traceable on a spur projecting from the southern face.

The west face is now practically composed of a great mass of Boulder Clay occupying the interval between the erratic and the Kimeridge Clay which forms the other bank of the valley.

Prof. Bonney, describing the eastern face, says: "Beginning at the north end the Lower Chalk, [which is] marly and shows traces of bedding distinctly dips at an angle of about 30° to N., the Upper [Cambridge] Greensand with phosphate nodules and characteristic fossils is in its usual place below, and the Gault (with many phosphatic nodules and fossils) below that; *and then the Boulder Clay distinctly dips under it at about the same angle.*" The upper part of this Boulder Clay showed contortions, and at the southern end was full of blocks of carstone which slipped down or were torn off the banks of Kimeridge Clay apparently during the accumulation of the Boulder Clay, for the banks were capped by an extension of the Ely outlier of Lower Greensand. This section is now indistinct. The southern side of the pit is covered with talus and overgrown with grass, but it probably consists of Kimeridge Clay overlaid by Boulder Clay with fragments of the Lower Greensand, so far as one can judge from the scattered blocks on the surface.

Mr Skertchly[1] states that he saw Boulder Clay underlying all the Cretaceous rocks and cropping out on all sides of the mass.

In the *Middle Pit* Mr Skertchly says "we again find the Chalk and Gault flanked by Boulder Clay which lies

---

[1] *Geol. of Fenland, Mem. Geol. Surv.* p. 236.

in a hollow in the Kimeridge Clay. The sections are obscure, and the beds can be determined now with difficulty. The great boulder does not extend into the *Lower Pit*, but the Boulder Clay reaches the river in a narrow band with Kimeridge Clay on each side."

Summing up we find that on all sides Boulder Clay is interposed between the great erratic and the Kimeridge Clay, that Boulder Clay dips below and underlies the mass of Cretaceous beds, and that the Boulder Clay and erratic lie in a valley excavated in the Kimeridge Clay. Moreover, the beds of Kimeridge Clay are bent down where they abut on the Boulder Clay and erratic, as if by the weight of the latter; the edges of the beds are cut off and slickensided apparently by the same mass; the Lower Greensand beds, which in the immediate vicinity rest in their normal position on the Kimeridge Clay, are here shattered and displaced as if they had slid or been forced down a slope of Kimeridge Clay; the Cretaceous beds of the erratic do not correspond in dip or strike with the Lower Greensand in the pit, but are placed unconformably on it and with unusual relations to it. Finally, any hypothesis of faulting demands a complicated series of movements which is scarcely credible in this area.

In short the only reasonable view is that the Cretaceous mass is a gigantic transported boulder brought from the Chalk area by the glacier which formed the Boulder Clay[1]. Examples of such huge ice-transported masses are by no means unknown from other areas, and many exceed the Ely mass in size. Prof. Judd[2] has described many in the Lincolnshire drift; an enormous

[1] The Gault has not been found in place within nearly 3 miles, and the nearest Chalk Marl is 5 miles distant.

[2] *Mem. Geol. Surv. Explan. Sheet* 64, p. 246.

one has recently been described at Catworth in Hunting-donshire[1]; a mass 800 ft. long and 60 ft. high is mentioned by Searles Wood[2]; in South Wales an isolated mass of sandstone 200 yds. long has lately been noticed by the Survey officers; and many others have been recorded on the Continent and in North America.

THE HIGH LEVEL GRAVELS AND LOAMS.

Resting generally on the table-lands and higher ground there are small patches of certain gravels and loams which have been termed 'Hill Gravels,' or the 'Coarse Gravel of the hills[3].' They probably mark the last stage of glacial conditions in this area. These 'High Level Gravels' appear to be the remains of a wide-spread deposit which has now been almost entirely removed by denudation. Probably the existing patches owe their preservation to lying in old channels or hollows in the rock beneath.

They are found resting on the Boulder Clay and overlapping it on to the Chalk (Fig. 11, p. 188). It has been suggested that some of those on the Chalk are of earlier date than the Boulder Clay[4]; and others may be only the Clay weathered *in situ*. Many geologists

---

[1] *Glacialists' Magazine*, vol. 1 (1893), No. 4, p. 96.

[2] *Geol. Mag.* Dec. I. vol. VIII. (1871), p. 409.

[3] Jukes Browne, *Post Tertiary Deposits of Cambs.* (Sedgwick Essay, 1876), p. 40.

[4] *Mem. Geol. Surv. Explan. Quart. Sheet* 51, S. W. p. 79.

hold that they are of marine origin and indicate a submergence and subsequent emergence of the land.

Characters. The gravels are composed of such materials as would be derived from the washing and sorting by water of the Boulder Clay itself, and this has led to the belief that they may be only rearranged Boulder Clay. Most of the pebbles are of flint, but there is an abundance of fragments of white chalk; and various Jurassic rocks and phosphatic nodules from the neighbourhood are also found in them. Of rocks from a distance fragments of Red Chalk, quartzites, Carboniferous Limestone and sandstone, with a variety of igneous and metamorphic rocks (basalts, porphyrites, granites, gneisses, schists, etc.) are fairly common.

The flints are usually angular but sometimes more or less rounded. The softer chalk pebbles are often well rounded. The pebbles of other rocks are generally subangular but the far-travelled boulders are much waterworn. Ice-grooves and scratches are occasionally preserved on them.

When the pebbles are imbedded in any matrix the latter is of a loamy, chalky, or sandy character.

The gravels themselves are for the most part coarse and totally unstratified with a considerable amount of iron-staining. Thin and impersistent beds of loam and sand are occasionally found intercalated.

Distribution. As above stated these gravels lie mostly on the high ground. The lowest point which they reach is in the old valley east of Babraham. East of Tharfield there is a loam on Boulder Clay on the top of the escarpment 500 ft. above the sea, which probably should be correlated with these gravels.

On Barrington Hill near Linton a patch of coarse flint-gravel lies directly on the Boulder Clay which caps the hill about 350 ft. above sea-level.

The large patch of gravel east of Hildersham belonging to the same group of deposits lies partly in a hollow of the Boulder Clay and partly on the Chalk. The junction is exposed in the cutting at the cross roads. The lower patch of sandy gravel north of Hildersham Church at the junction of roads and lying about 140 ft. above sea-level was probably once connected with that on the hill above. Westwards the Boulder Clay is overlaid by thick beds of gravel which extend beyond the Boulder Clay on to the chalk.

A pit ¼ mile north-west of Little Abington Grange shows gravel composed of pebbles of a great variety of rocks. Again, in the pits north of Fulbourn Lodge 6 to 8 feet of sand and chalky gravel may be seen resting on the Chalk.

Copley Hill, Misleton Hill, and Little Trees Hill are capped by chalky gravel with many flints.

The old pits ½ mile west of Hill's Farm on the north side of the old Roman road called Worsted Street have been described by Prof. Sedgwick and others. According to the *Geol. Surv. Memoir* the beds exposed in 1875 were as follows :—

Stiff yellowish sandy clay, containing stones of various sizes and set at various inclinations; the pebbles of chalk rounded, the others angular. About 50 per cent. are chalk, 30 per cent. flints and 20 per cent. of various other rocks ...   6 ft.

Coarse rubbly gravel, said to be obtained below this and resting on chalk ...   . 3–4 ft.

The phosphatic nodules from the Cambridge Green-sand, which are found in this locality in the gravel, are more than 200 ft. above the level of the present outcrop of their parent bed lying to the north and north-west.

There are many small patches of gravel on the Gog Magog Hills belonging to this series, as for instance on the slope from the escarpment towards Newmarket. Loamy patches occur a mile south-east of Westley Waterless.

The ridge of Boulder Clay N.W. of the Ouse is capped by patches of similar gravel, as for instance near Bluntisham and Pidley.

Origin. It has long been held that these gravels were formed by the washing out and rearrangement of the Boulder Clay[1]; the materials, the condition of the pebbles, and the distribution of the deposits all point to the same conclusion. But whether this process was connected with the last stages of the Glacial Period or belongs to more recent times is a disputed point. It is probable that they correspond with the "deposits of stratified flinty or chalky gravel or sand" which overlie the Chalky Boulder Clay of the Midlands and which, according to Prof. James Geikie[2], are the so-called "Middle Sands" etc. of Cheshire. It is supposed by the same writer that when the glaciers commenced to retreat great volumes of water from the melting of the ice rushed over the land in a torrential manner, washing out and rearranging the coarser materials of the Boulder Clay and carrying off the finer particles in suspension to be deposited in quiet inlets as beds of sand or loam. It seems, however, very questionable whether such floods can explain their present distribution.

[1] Seeley, *Q. J. G. S.* vol. xxii. (1866), p. 479. Bonney, *Camb. Geol.* p. 52.

[2] Geikie, *Great Ice Age*, 3rd ed. (1894), p. 376.

It is urged, moreover, by Prof. Geikie that as the North Sea ice-sheet had advanced against the drainage of the country the larger streams on its retreat flowed down again into their clay-filled valleys, but because their waters could for a long time find no escape into the open sea owing to the opposing front of the ice-sheet the lowlands must have been submerged by the formation of a large lake or series of lakes, such as Prof. Carvill Lewis supposed[1]. The size and depth of these lakes depended on the contour of the ground, the height of the ice-front, and the number and magnitude of the streams obstructed. The lakes were not completely drained off till the final disappearance of the ice-sheet. It is supposed that the draining at last took place suddenly, as was once the case in the modern Merjelen Lake[2], so that the water rushed in a torrential manner over the land, washing out the soft clay, roughly sorting its contents and heaping them together in masses of unstratified clayey and chalky gravel. On this theory it is difficult to understand their presence on the hills and absence in the valleys into which we might have expected the materials to have been swept, but it is held that in the valleys they have been incorporated into the river gravels, while post-glacial denudation has spared them only in sheltered situations on the crests of ridges.

Those geologists who believe that the Chalky Boulder Clay is a submarine deposit formed of material brought by icebergs explain these High Level Gravels as the result of wave-action on the surface of the Boulder Clay as the land emerged from the sea[3]. If we deny that they

[1] *Glacial Geology of Great Britain and Ireland*, p. 42.
[2] C. S. Du Riche Preller, *Geol. Mag.* Dec. 4, vol. III. 1896, p. 97.
[3] W. H. Penning, *Q. J. G. S.* vol. XXXII. (1876), p. 191.

are simply the result of the post-glacial weathering *in situ* of the Boulder Clay, and also maintain that there is no direct evidence of their formation by marine agency, or of the submergence required, we must have recourse to some theory which calls in the action of land-water on an extensive scale. However no explanation hitherto offered is wholly satisfactory.

POST-GLACIAL DEPOSITS.

THE RIVER GRAVELS[1].

The river gravels in the Cambridge valley belong to two main groups: (1) the gravels of the ancient river-system; and (2) the gravels of the present river-system. The latter gravels occur in three main terraces which are termed (1) the Barnwell terrace; (2) the intermediate terrace; and (3) the lowest terrace (Fig. 11, p. 188).

Between the deposition of the gravels of the ancient river-system and those of the present river-system there must have been an entire change in the drainage of the area, for the channels of the present rivers and streams rarely coincide with the ancient ones. What led to this widespread change was undoubtedly earth-movement of some kind which diverted the streams into new courses and thus led to the formation of new valleys and surface-features.

Prior to the formation of any of the river gravels[2] we find incontestable evidence in the pre-glacial valleys descending far below the present sea-level (see p. 161) that there was an important downward movement of the land leading to submergence of the low-lying coastal region and the lower parts of the river-valleys. If we hold the

[1] Jukes Browne, *The Post Tertiary Deposits of Cambridgeshire* (Sedgwick Essay, 1876). *Mem. Geol. Survey, Quart. Sheet* 51, S.W. pp. 82—111. *Mem. Geol. Survey, Quart. Sheet* 51, N.E. pp. 72—81. *Mem. Geol. Survey, Sheet* 65, pp. 90—104.

[2] Mr Searles V. Wood believes that the ancient gravels are not due to river action, and maintains that they were deposited in extra-morainic lakes on the retreat of the ice (*Q. J. G. S.* vol. xxxviii. (1882), p. 667).

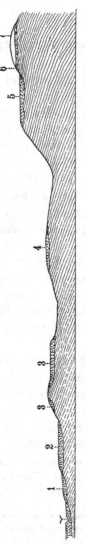

FIG. 11. DIAGRAMMATIC SECTION OF THE GRAVELS BETWEEN BARNWELL AND LINTON.
(By Prof. T. MᶜKenny Hughes.)

⅄ Present alluvium of river.
1 The Lowest terrace (Chesterton Common).
2 The Intermediate terrace (Fen Ditton and Chesterton).
3 The Barnwell terrace.
4 The Gravel of the Ancient River System.
5 The High Level Gravel above and derived from the Boulder Clay (Wandlebury Gravel).
6 The Linton pebbly Gravel (below the Boulder Clay).
7 Boulder Clay.

view that the Boulder Clay and High Level Gravels owe their origin to coastal ice, ice-bergs or marine agencies, this submergence must have taken place at the beginning of or during the Glacial Period. This view also necessitates that between the formation of the Forest Bed Series and that of the Arctic Freshwater Bed there was a long interval, during which the whole of East Anglia and probably the whole of Britain was raised far above the level of the sea[1]; for it was in this period of elevation that those valleys were cut out which now have their floor far below sea-level. According to this theory the submergence which took place about the beginning of Glacial times depressed the land which occupied the area of the North Sea more or less completely beneath the water, but after the formation of the Glacial Drift there came an uplift sufficient to re-unite England with the Continent, and to allow of the immigration of large mammals, etc. belonging to the continental fauna. The land however did not attain its former height, but during this second period of elevation the earlier river gravels of this district, including the Barnwell Terrace, and the local moraines of mountainous districts further north were formed. Some slight earth-movement—perhaps only of local importance—caused the shifting of the courses of the rivers shown by the difference of the trend of the ancient river gravels and of those of the Barnwell Terrace. Some time after the gravels of the Barnwell Terrace were formed the land connection with the Continent across the English Channel was finally broken, but there seems to be no evidence that this movement affected our area to any important extent.

On the other hand, if we hold the land-ice theory of

[1] Jukes Browne, *Building of the British Isles*, p. 286.

the origin of the Glacial Drift, we do not require so many repeated earth-movements. According to this theory there was upheaval at the close of the Pliocene period of the whole of Britain, during which time were cut out the deep valleys which are now below sea-level. Glacial conditions then set in, probably before the land had reached its greatest altitude. A thick covering of ice swathed the mountainous regions and traversed the lower grounds in huge confluent glaciers or ice-sheets. After this icy mantle had spread over the greater part of Great Britain, and probably twice over East Anglia, it commenced to melt and retreat from the lowlands. Unequal depression of the land accompanied or followed this departure of the ice, bringing the floor of some of the old valleys below sea-level; and on the drift-covered surface of the land-area the rivers of the ancient system arose, cut out their valleys, and deposited their gravels. A slight modification in the downward movement or a warping of the surface initiated the present river-system. During the formation of the Barnwell gravels and perhaps later, England maintained a con-tinental connection, and probably possessed a consider-able eastward extension over the North Sea. Further differential movements followed, mainly tending to sub-mergence; the land connection with the Continent was destroyed shortly before or very soon after the advent of Neolithic man; and the latest movements in East Anglia are probably indicated by the submerged forests on the coast.

There is however considerable difference of opinion about these later Pleistocene movements, but it would be beyond the scope of this work to enter more fully into the question[1].

[1] For further details see the Survey Memoirs of the district, Prof.

Economics. A considerable amount of water is obtained from the gravels by wells sunk through them to the surface of the Gault, but the quality of the water is usually bad. The gravels also supply the district with abundant road-metal.

THE GRAVELS OF THE ANCIENT RIVER-SYSTEM.

These early river-gravels, which are found along the valleys running northward and north-westward through the Chalk hills and over the lower ground at the base of the latter, somewhat resemble the coarse High Level Gravels in composition, but they contain a much smaller percentage of far-transported material and show more evidence of rolling. They may be distinguished also by their distribution and to a minor extent by the presence of more or less distinct bedding and of subsidiary beds of sand and loam. The High Level Gravels occur usually as isolated outliers or patches, are limited for the most part to the higher ground, and do not appear to bear any relation to the direction of ancient or modern valleys in the Chalk hills. The ancient river-gravels on the other hand are found at lower levels (Fig. 11, p. 188) and occur as series of elongated patches along the sides or bottom of the long but now often dry valleys which cut through the Chalk escarpment. As they descend into the plain the patches become larger, and those from adjacent valleys tend to unite to form long ridges of gravel extending in definite lines across the country. Two considerable streams which have been

Boyd Dawkins' *Early Man in Britain*, Wallace's *Island Life*, p. 319 *et seq.* Woodward's *Geology of England and Wales*, p. 481, etc. Prof. T. McK. Hughes, "Evidence of Later Movements of Elevation and Depression," *Proc. Vict. Instit.* vol. xiv. 1880—1881. Prof. J. Geikie's *Great Ice Age*, 3rd ed. etc.

traced by means of these gravel-patches appear to have met in the neighbourhood of the town of Cambridge, whence they flowed in a north-north-westerly direction past Chatteris to March, where the river entered the sea. The details of the course of this ancient river with its tributaries and affluents are given below.

The fluviatile origin of these deposits is proved by their distribution along and relation to the upland valleys, by their behaviour on the lower ground, and by the occasional occurrence in them of land and freshwater shells and the bones of terrestrial mammals.

The great age of these deposits is indicated by their want of dependence on the present surface-features and by their divergence from the existing lines of drainage. In places the later fluviatile deposits and the existing rivers cross them at right angles.

Characters. There are no persistent lithological or structural characters in these ancient river-deposits. Gravels, sands, and loams all enter into their composition. The gravels usually contain a large amount of chalky material. The pebbles are mostly of chalk and flint, but there are also many of quartzite, sandstone, limestone, schist, and igneous rocks,—derived from the Boulder Clay or High Level Gravels. In some places the proportion of chalk stones is more than 50 per cent.

There is usually some kind of bedding present, and in the finer gravels false-bedding is not uncommon.

The sands and loams are inter-bedded with the gravels in thin inconstant patches, and sometimes show contortion, as at Whittlesford. Occasionally the sands and loams preponderate and form beds several feet thick. False-bedding in them is common, and they frequently cut into the gravels or the gravels into them, in troughs or channels. The beds are often stained with iron.

Distribution. The courses of the ancient rivers are indicated by these patches of gravels, but it is not always easy to separate them from the more recent river-deposits, particularly when the ancient river-course coincides with the recent one.

Beginning in the south-western part of the district we find a series of gravelly and loamy patches starting from Wardington Bottom near Royston and running along a north-easterly line to Whittlesford. The course of the modern little stream coincides with them only for a short distance and bends away north-westward to flow past Foulmire. The ancient channel is traced by outliers of gravel, sand, and loam on the slope of the hills by North Hall, Sharpens Farm, Heydon Grange, north of Chrishall Grange, towards Triplow and then eastward to Whittlesford, where in a gravel-pit close to the station a bed of false-bedded fine gravel is overlaid and cut into by sandy buff-coloured loam, above which occurs a second bed of gravel which has cut out a channel in the loam. At Whittlesford this ancient stream from Wardington Bottom appears to have been joined by two others, one of which flowed along the present Cam or Granta valley from Quendon in Essex, and the other ran approximately along the modern valley of the Lin. The predecessor of the modern Granta no doubt left deposits similar to those of the Wardington stream, but they have been so mixed up and incorporated with the modern river gravels lying in the same valley that it is almost impossible to separate them. Probably some of the loams near Wenden and Newport belong to the ancient river deposits, and possibly either the loam or gravel or both

R.   13

at Chesterford should be ascribed to the same series, but this is doubtful.

The eastern affluent—the predecessor of the modern River Lin—ran along the present valley, but at a higher level than the existing river. Patches of the ancient gravel with thin bands of sand and loam are seen in several pits near Bartlow Station and appear banked up against the Chalk slope. Further down the stream on either side of Linton and again near Pampisford Station and Abington Park to the south of the present river, patches of gravel seem to lead on to the Whittlesford mass.

The subsequent course of the river formed by the union of the three streams at Whittlesford cannot be traced with certainty, since no newer ground high enough to reach its level is now found on the south side of Cambridge. The river, however, must have flowed through the gap between the Gog Magog Hills on the east and the Cretaceous promontories of Haslingfield and Barton on the west. The small patch of gravel near Stanmoor Hall may be a remnant of the deposits of this vanished stream, while the gravel cap on Redland Hill near Harston may point to the existence of another tributary. But all further evidence of the course of the river towards Cambridge has been denuded away or obscured by more recent deposits.

A completely distinct system of drainage existed on the north side of the watershed above Hildersham, and reference must first be made to it before tracing the course of the main river north of the town of Cambridge. From the hills about Hildersham and Balsham several deep and long valleys run northwards and north-west-wards and contain numerous patches of gravel and sand belonging to the ancient river system. These valleys are now for the most part dry. One such valley runs in a north-westerly direction from New Yole Farm south of Balsham, and a series of patches of gravel, sand, and loam are seen in it at various intervals and in pits near Balsham and Dungate. At the latter place the patch lies on the slope of one side of the valley only, for the opposite side has been worn away more rapidly by the deepening and lateral corrosion of the valley in subsequent times. This is not an uncommon feature along these ancient valleys, but where they widen out into the lower and more open ground the old gravels usually form long ridges elevated above the present general level of the land. The enormous amount of denudation that has taken place since their formation is forcibly impressed upon us by these features. The railway line to Newmarket cuts through one of the ridges along this Balsham line of gravels, and beyond this point the ridge is prolonged till cut off by the subsequently ex-cavated hollow in which Great Wilbraham lies. On the further side of this hollow the long gravel-capped ridge to Quy-cum-Stow is plainly a prolongation of the same.

To the north-east of this Balsham valley lie two other valleys running north-westward from the slopes near West Wratting and Weston Colville. Both of these contain a similar series of gravels. In the first of these valleys the highest patch of gravel is very small and occurs 1 mile north of Balsham Church. Further down the valley a small pocket of gravel in a hollow near Wadley Hall has yielded many mammalian bones. So has also a larger patch near Lark's Hall.

In the second valley only small patches are found, but they occur along a line which joins the gravels of the other valley near Six Mile Bottom, and this union leads

13—2

to a spread of gravel. The two streams which here seem to have become confluent, flowed as one river north-westward to Wilbraham, the course being now indicated by a long ridge of gravel; the sides of the valley have been removed by denudation. Near Wilbraham it was joined by the stream from the westernmost of this group of three valleys, and the united streams flowed on westwards. The gravels, marking the further course of this waterway lie on the top of a long ridge of Chalk which they have preserved from denudation and which corresponds to the bottom of the ancient valley; the hill-slopes which once formed its sides have been entirely washed away. We are thus again enabled to get some idea of the enormous change in the configuration of the country since these gravels were deposited. Pits in this ridge have been opened near Little and Great Wilbraham, and freshwater shells and mammalian bones have been recorded from them by the Geological Survey. Beds of loam and sand are exposed in some of the pits.

This ridge is continued to Quy-cum-Stow, beyond which it is interrupted, but patches of gravel along the Newmarket road with the same trend plainly carry it on to Fen Ditton where it is cut across by the present valley of the Cam. Most probably the gravels on the Chalk Marl ridge on the opposite side of the valley, *i.e.* those on Castle Hill and Mount Pleasant, are a prolongation of the same series, while just beyond the Observatory the large pits by Gravel Hill Farm are in gravel, presumably belonging to the same. The gravel here is about 5 feet higher than that at Quy-cum-Stow, which apparently is due to a slightly unequal upheaval of the district at some subsequent time—probably at the time when the modern system of rivers began to flow.

It was somewhere close to this spot that the river from the east, just described, was joined by that which flowed from the south which apparently occupied the present valley of the Cam between Harston and Cambridge.

From the Observatory the line of gravels trends north-north-west towards Girton, for the stream from the south being more powerful owing to its greater volume, which depended on its larger catchment-area and greater length, prevailed over its eastern affluent and caused the river that resulted from the union to pursue a more northerly course.

The gravels in passing from the Observatory to Girton underlie St Giles' Cemetery and cross the Huntingdon Road. The ground on which they occur forms a perceptible ridge. Girton College stands on this gravel, which is here more than 10 feet thick. Between Girton and Histon the ridge trends slightly north-east; thence north-west to Oakington, throwing out springs at its base on its western side, and somewhat confused owing to breaches by small streams which have re-deposited the gravel and mixed it with later deposits. Beyond Oakington the high ridge sets in again and is continued steadily past Long Stanton, across the railway that goes to St Ives, and on to Swavesey and Over. Here as elsewhere this old river-gravel occupies high ground and has no relation to the existing valleys. Perhaps the large outlier of gravel around Willingham belongs to this series.

About Bluntisham and Colne gravel also occurs and may belong to this ancient line of drainage. Further north at Somersham there is a large patch beneath the town and on the high ground to the west and north-west. Chatteris is situated on a small island of gravel and loam

rising above the level of the surrounding fens, and marine
shells have been found here[1]. Following this line of
gravels still further to the north we come to the patch
of gravel around Doddington and Wimblington, from
which many marine shells have been obtained. A short
distance further north is the important gravel of March,
which has been the subject of considerable discussion[2], but
there seems but little doubt that it belongs to the series
of gravels here described and is not "contained between
Boulder Clays" as Prof. Seeley held[3]. There are several
patches of gravel around March which rise out of the
fens as an island nearly 7 miles long. The gravel varies
in thickness from 6 to 20 feet, and rests in some places
on Boulder Clay and in others on Kimeridge Clay. It
is rudely stratified and contains abundant chalk pebbles;
its characters vary however considerably within short
distances. A brownish clay or loam overlies the gravel
here and there; this was held by Prof. Seeley to be
Boulder Clay and to prove the interglacial age of the
gravels. It is possible that though the March gravel
belongs to our ancient river system of gravels and is
newer than any glacial bed in the district, yet it may
be older than some of the glacial beds in the north of
England. Mr Clement Reid correlates the March gravels
with those of the Holderness drift on the evidence of the
fauna[4], with which Prof. Hughes has compared the shell-
bearing drift of the Vale of Clwyd[5]. The gravel patches

[1] *Mem. Geol. Surv. Geol. of Fenland*, p. 202.

[2] *Mem. Geol. Serv. Explan. Sheet* 65, p. 105, for details and
references.

[3] Seeley, *Q. J. G. S.* vol. xxii. (1866) pp. 472, 473, 480. *Geol. Mag.*
vol. iii. (1866) pp. 500, 501.

[4] *Mem. Geol. Surv. Geol. of Holderness*, pp. 68—71.

[5] *Q. J. G. S.* vol. xliii. (1887) p. 73.

at Eastrey, Coates, Whittlesea, and around and north of Peterborough are of the same age, and are to be regarded as the marine representatives of the ancient river-gravels which pass into them. They all yield marine shells in abundance, and occasionally freshwater shells are intermixed. Eastrey and Whittlesea rise above the fen as islands, being formed of bosses of Oxford Clay capped by gravel. Numerous pits are worked in these localities, in which the gravel is seen to attain a thickness of from 8 to 12 feet[1].

These marine shell-bearing gravels appear to have been formed at the mouths of the rivers which emptied themselves into the ancient Wash when its shore-line was situated much further south and west than it is now. The mixture of marine and freshwater shells is thus explained, but the preponderance of the former shows that marine rather than estuarine conditions prevailed. The most abundant species of the molluscs live at a depth of from 5 to 10 fathoms, but the shells of those found in the gravels were thrown up by the waves on the beach which the gravels formed.

The marine gravel at Hunstanton[2] may be of the same age.

Returning now to the ancient river-gravels which are found in the valleys, we find in the neighbourhood of Newmarket several patches of gravel arranged along

---

[1] *Mem. Geol. Surv. Geol. of Fenland*, pp. 187—192.

[2] Rose, *Lond. and Edinb. Phil. Mag.* 3rd ser. vol. VIII. (1836), p. 34. Th. Wiltshire, *Proc. Geol. Assoc.* vol. I. (1859), pp. 8—11. H. G. Seeley, *Q. J. G. S.* vol. XXII. (1866), p. 470. S. V. Wood and Harmer, *Suppl. Crag Mollusca (Palæont. Soc.)*, p. 228. Jukes-Browne, *Q. J. G. S.* vol. XXXV. (1878), p. 415. B. B. Woodward, *Proc. Geol. Assoc.* vol. VIII. (1883), p. 97.

definite lines, but apparently belonging to a different river-system to that above described. These patches commence in the valleys which run north and north-westward from near Westley Waterless, Dullingham, Ditton, and Cheveley. They seem to converge near Exning. Some other patches on Ling Hill south-west of Newmarket may indicate a tributary. From Exning the ridge of gravel turns to the north-east and seems to be connected with the patch of gravel at Snailwell on the Newmarket and Ely railway line and with the long narrow ridge of gravel stretching south of Landwade. There is chalky gravel in pits near Kentford and Kennet, and it may be of the same age as the patches above mentioned.

These patches probably belong also to the same series as those on Barton Hill, south of Mildenhall, and near Eriswell and Lakenheath further north[1]. They "follow a line almost at right angles to the present rivers, and they extend right across the Kennett, the Lark, and the Ouse, so that they are clearly independent of the present rivers" (*Skertchly*). Flint implements have been found in several of these localities (see p. 238). The gravels possess the usual characters of those of the ancient river system, and at Snailwell the basement bed which rests on the Chalk is of sand.

North of Lakenheath the course of these gravels has not at present been determined.

It must be remembered in examining the gravels of the dry chalk valleys that they work down by subterranean denudation so that their base and basement bed have often been modified after the accumulation of the main mass of gravel.

[1] *Mem. Geol. Surv. Explan. Sheet* 51, N.E. pp. 72, 73.

LIST OF FOSSILS FROM THE ANCIENT RIVER GRAVELS.

MAMMALIA.

> *Bos primigenius* Bojan.
> *Elephas antiquus* Falc.
> *Equus fossilis* H. V. Meyer
> *Hippopotamus major* Desm. (=*H. amphibius* Linn.)
> *Rhinoceros tichorhinus* Cuv.

MOLLUSCA.

> *Helix hispida* Linn.
> *Pupa marginata* Drap. ?
> *Succinea putris* var. minor
> *Pisidium amnicum* Müll. ?

LIST OF FOSSILS FROM THE MARCH GRAVELS.

(The commonest species are marked with an asterisk ✱.)

CRUSTACEA.

> *Balanus porcatus* Da Costa

BRACHIOPODA.

> *Rhynchonella psittacea* Chemn.

MOLLUSCA.

Lamellibranchiata.

> *Anomia ephippium* Linn.
> *Astarte borealis* Chemn.
>   „   „   striated variety
>   „   *compressa* Mont.
>   „   *sulcata* Da Costa
> ✱*Cardium edule* Linn.
> *Corbicula (Cyrena) fluminalis* Müll. ?
> *Corbula gibba* Olivi (=*C. striata* Lam. and *C. nucleus* Lam.)
> *Cyprina islandica* Linn.
> *Mactra solida* Linn. (=*M. ovalis* Sow.)
> *Mya arenaria* Linn.
>   „   *truncata* Linn.

Mollusca (*cont.*).

Lamellibranchiata (*cont.*).

Mytilus edulis Linn.
  „   *modiolus* Linn.
Ostrea edulis Linn.
Pholas crispata Linn.
  „   *dactylus* Linn. ?
Scrobicularia plana Da Costa
Tellina balthica Linn.
  „   *calcarea* Chemn. (= *T. lata* Gmel.)
Unio tumidus Retzius

Gasteropoda.

Aporrhais pes-pelicani Linn.
Buccinum undatum Linn.
Bithinia tentaculata Linn.
Emarginula fissura Linn.
Hydrobia (*Rissoa*) *ulvae* Penn.
Lacuna crassior Mont.
  „   *vincta* Mont.
Littorina littorea Linn.
  „   *rudis* Maton
Natica alderi Forbes
  „   *catena* Da Costa
  „   *islandica* Gmel. (= *N. helicoides* Johnst.)
Pleurotoma pyramidalis Ström.
  „   *rufa* Mont.
  „   *turricula* Mont.
Purpura lapillus Linn. (including var. *imbricata*)
Scalaria communis Lam.
Trochus cinerarius Linn.
Trophon bamffius Mont.
 *   „   *clathratus* Linn.
Turritella terebra Sow. (= *T. communis* Risso)
Valvata piscinalis Müll.
Velutina undata J. Smith

Scaphopoda.

Dentalium entalis Linn.

GRAVELS, ETC. OF THE PRESENT RIVER SYSTEM.

The gravels which lie more or less along the courses of the present rivers, following in the main the present lines of drainage, and occurring in the valleys and along the banks of the existing streams, have now to be considered. They are however found at different levels to the deposits of the rivers whose waters now flow across the country, and do not resemble them in character, for in historical times only silt and sand or fine gravel have been deposited, whereas the coarseness of these earlier gravels as well as their thickness and extent indicate a greater carrying power, a larger volume of water, and less temperate conditions than now prevail.

In the upper parts of the valleys extensive denudation has usually destroyed the gravel terraces, washing the material down to lower levels and incorporating it into river alluvium of a later date. The lateral denudation has not been so great on the gentler slopes of the lower portions of the valleys, so that it is here that we find the gravel terraces most distinct and best preserved. Such is the case just south of Cambridge and in the neighbourhood of the town. On the more open and level country to the north there is scarcely any difference of level between the deposits of various periods, and consequently the terraces are very difficult to separate.

In the Cam River-system we can distinguish three terraces. The oldest and highest of these is known as the *Barnwell Terrace* because of the good sections of it at that place; the middle one is termed the *Intermediate Terrace*; and the newest one is known as the *Lowest Terrace*.

FIG. 12. SECTION IN BRICK-PIT BETWEEN BARNWELL STATION AND STOURBRIDGE COMMON[1].

(Reproduced by kind permission of Prof. T. McKenny Hughes and the Editor of the *Geological Magazine*.)

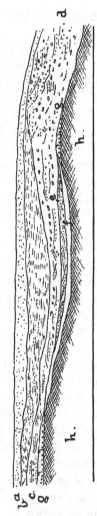

Scale : 20 feet to 1 inch.

a. Rusty sand.
b. Buff sandy loam, darker where damp, some cross-bedding, lines of more sandy loam weathering out.
c. Gravel, split up by beds of sand and loam at S.E. end.
d. Gravel, falling over bank to lower level of Stourbridge Common.
e. White sand and gravel with shells, which runs higher further west.
f. White marl, such as might be derived from the Chalk Marl, and bands and pockets of re-sorted phosphatic nodules.
g. Remains of Chalk Marl and Cambridge Greensand running into *f*.
h. Gault.

[1] *Geol. Mag.* Dec. 3, vol. v. (1888) p. 195, fig. 2.

(a) *The Barnwell Terrace.*

The gravelly silt at Barrington[1] is the southernmost patch of any importance belonging to this terrace. It marks the spot where the river that rose in Wardington Bottom was forced to turn to the north-east on account of the opposing barrier of the Chalk hills near Orwell and Barrington. We have seen that when the ancient river-system prevailed the stream from Wardington Bottom flowed from its source in a north-easterly direction; but it seems to have gradually changed its course, first to the north and then to the north-west from Foulmire to Foxton. The patches of gravel near Foulmire, Foxton, and Harston as well as the gravelly silt near Barrington on the opposite side of the present Rhee valley result from the obstruction by the western Chalk hills of the waters of this Wardington river. For a large swampy area, liable to frequent floods so as to form a lake, would originate at this spot, and on its southern side where the river entered it would be dropped all the coarse material from the hills to the south, while the finer material would be carried out further into the lake and deposited under the lee of the Chalk hills, down whose slopes would be washed much marly clay from the denudation of the Chalk and of the Boulder Clay. Thus at the south-eastern corner of the flooded alluvial flats a set of beds would be accumulated completely different in character to those on the western side. The latter are the Barrington silts and marls; the former are perhaps the Foxton gravels. It is probable that there was no Rhee valley above Barrington at that time, but that the whole

---

[1] Rev. O. Fisher, *Q. J. G. S.* vol. xxxv. 1879, p. 670. Mrs T. McK. Hughes, *Geol. Mag.* Dec. 3, vol. v. (1888), p. 193.

country was occupied by Chalk hills capped with Boulder Clay; for the present Rhee is entirely fed by springs from the Chalk Marl, the outcrop of which must then have occupied a very different position to what it does now. The Rhee valley above Barrington is marked by an absence of gravelly deposits, and this points to the removal of the Boulder Clay from the surface of the land prior to the excavation of the valley, for the gravels elsewhere derive most of their materials from the Boulder Clay.

In this shallow lake at Barrington the carcases of numerous large mammals were imbedded in the silt. The great abundance of their remains must be due to a peculiar combination of circumstances. No doubt the marsh was the haunt of the hippopotamus, and the favourite drinking resort of herds of ruminants and pachyderms, while the hyænas, bears, and lions roamed round its banks to prey on the herbivores. Perhaps an eddy in the sheltered western corner of the lake prevented the carcases drifting away[1]. Land and fluviatile shells are abundant in this silt, and many of them as well as a large number of the mammalia are found also at Barnwell (see list p. 210).

From the neighbouring valleys of the Cam and the Lin issued streams which after uniting joined the Wardington river above Harston, but the oldest terrace in these tributary valleys can rarely be identified with certainty.

All along the upper part of the Cam valley in N.E. Essex patches of gravel occur in which shells and mammalian bones have been found. Such patches occur at

---

[1] Rev. O. Fisher, *Q. J. G. S.* vol. xxxv. 1879, p. 670.

Newport, Wenden, and Little Chesterford, but the different terraces have not so far been separated[1].

Lower down the Lin valley patches of gravel occur which apparently represent the Barnwell terrace, and north of Pampisford there are 10 ft. of stratified gravel and loam exposed in some pits containing *Limnœa* and *Pupa*.

It is just beyond the old point of junction of the Lin and Cam and on the high ground N.W. of Shelford that we get a patch of gravel which can with certainty be ascribed to the Barnwell terrace. The railway cuts through it, and the deposit is seen to consist of fine gravel, white sand, and marl, resting on the Chalk slope. It contains land and fluviatile shells including the characteristic *Cyrena* (*Corbicula*) *fluminalis*. Other patches occur between this spot and Trumpington, where the gravel still rests against a Chalk slope, as the river had not at that time cut down to the Gault.

Near Trumpington the main stream received a tributary from Comberton and Barton. Patches of gravel at Comberton and between that village and Barton lead on to the western end of the long ridge of gravel which extends continuously from Barton to Grantchester, where it is interrupted by later gravels and the modern Cam. The stream known as the River Bourn flows to the south of the Barton gravel ridge.

North of Trumpington the river of the Barnwell Terrace appears to have turned towards the north-east, but its deposits are here again cut through by a set of later gravels trending north-west. Close to the Cambridge Railway Station the gravel again sets in and reaches a thickness of about 16 feet. It runs then north-west to

[1] *Mem. Geol. Surv. Explan. Sheet* 47, pp. 70, 71.

Mill Road, underlies the cemetery, extends to the western corner of Parker's Piece, along East Road, and thus to Barnwell, where in the pits near the Abbey Church bedded marls, sands, and gravels reach a thickness of 7 to 20 feet. To the north-west they overlap the Chalk Marl and come to rest on the Gault.

Bones and teeth of *Rhinoceros tichorhinus*, *Elephas primigenius* and *Equus fossilis* have been recorded from several of the above-mentioned patches of gravel. But at Barnwell—the typical locality for the development of this terrace—a much more abundant fauna has been found[1] (see list). According to the Survey Memoir (*Explan. Quart. Sheet* 51, S.W. p. 98) the general succession seems to have been the following :—

		feet
(6)	Soil and disturbed gravel	3—4
(5)	False-bedded sands	⎫
(4)	Layer of white marly loam	⎬ ... ... 10—14
(3)	False-bedded sand and fine gravel	⎭
(2)	Brown marly clay or loam	½—4
(1)	Coarse pebbly gravel	2—4
	Total	about 20 feet

Flint pebbles are the chief component of the gravels here. Chalk pebbles and phosphatic nodules are also common. Boulders of far-travelled rocks, such as gneiss, granite, basalt, felstone, quartzite, limestone, etc. also occur. Most or all of these have been derived from older gravels which had originated from the denudation of the Boulder Clay.

Beyond Barnwell the Barnwell Terrace strikes across the present river valley past Chesterton, but it is then cut out by the later gravels. Between the windmill N.W.

---

[1] T. McKenny Hughes, *Geol. Mag.* Dec. 2, vol. x. (1883) No. 10, p. 454. Mrs T. McKenny Hughes, *Geol. Mag.* Dec. 3, vol. v. (1888) p. 193.

of Chesterton, which stands on an island of Chalk and
Gault, and Castle Hill there extends a trough nearly half
a mile wide filled with gravel and loam, and at the south
end of Victoria and Milton Roads a pit is opened in
sandy brickearth and gravel; and in it have been found
bones of *Cervus giganteus*, teeth of *Elephas primigenius*,
Rhinoceros tichorhinus, *Hippopotamus amphibius*, and
Equus fossilis, with the common shells *Cyrena* (*Corbicula*)
fluminalis, *Unio pictorum*, *Pisidium amnicum*, *Planorbis*,
Valvata, *Helix*, etc. Northward of this channel the
gravels spread out over a wide area towards Impington
and Histon, but the patches have suffered much denudation and rearrangement, so that it is difficult to be sure
of their age.

The patch of gravel north-east of Histon is cut off
quite suddenly along a line running east and west about a
mile north of the village, and there is an abrupt step-like
descent on to the Gault plain which stretches north.
There are no gravel patches of this age further north.
It appears probable that the river, finding no outlet to
the west or north-west on account of the high ground
south of Oakington which is capped with gravel belonging
to the ancient river-system, and having also no means
of escape to the north on account of the ridge of
Lower Greensand and Kimeridge Clay from Oakington
to Cottenham, was forced to turn north-eastwards. At
the elbow thus formed in its course it expanded into a
broad sheet of water in which the spread of gravel about
Impington and Histon accumulated.

The further course of the river appears to have been
north-eastwards, and the higher patches of gravel at and
beyond Denny Abbey seem to mark its bed. Beyond
the gravel outlier at High Elm House, N.E. of Causeway

End Farm, we are unable to trace its continuation, owing apparently to its removal by denudation or its burial beneath the fen deposits of peat and silt.

The following list of the fossils is compiled from the Survey Memoir, and the papers by Mrs M^cK. Hughes[1], and Mr B. B. Woodward[2]:

LIST OF FOSSILS FROM THE BARNWELL TERRACE.

* Barrington. † Barnwell. ‡ Milton Road and Chesterton. § Grantchester.

VERTEBRATA.

MAMMALIA.

      *†*Bison priscus* Bojan.
      **Bison* or *Bos* sp.
      *†§*Bos primigenius* Bojan.
      *†‡*Cervus giganteus* Blum. (= *C. megaceros* Hart.)
      *†§ ,,   *elaphus* Linn.
      *†*Elephas antiquus* Falc.
      *†‡§ ,,   *primigenius* Blum.
      *†‡§*Equus fossilis* H. V. Meyer
      *†§*Felis spelæa* Goldf.
      *†‡*Hippopotamus major* Desm. (= *H. amphibius* Linn.)
      *†§*Hyæna spelæa* Goldf.
      *†*Meles taxus* Linn.
      *†‡*Rhinoceros tichorhinus* Cuv. (= *R. antiquitatis* Blum.)
      †*Sus scrofa* Linn.
      *†§*Ursus spelæus* Blumb.
      †§*Arvicola agrestis* Linn. ?

AVES.

      Anser sp.

---

[1] *Geol. Mag.* Dec. III. vol. v. (1888), p. 193.
[2] *Proc. Geol. Assoc.* vol. x. (1888), p. 355.

INVERTEBRATA.

Mollusca.

Gasteropoda.

†§*Ancylus fluviatilis* Müll.

†§   „   *lacustris* Linn.

†§*Azeca tridens* Riet.

  †*Balea perversa* Linn. ?

*†§*Bithinia tentaculata* Linn.

  †*Bulimus lackhamensis* Mont.

†§*Buliminus montanus* Drap.

†§   „   *obscurus* Müll.

†§*Cæcilianella acicula* Müll.

†§*Carychium minimum* Müll.

†§*Clausilia pumila* Ziegler

†§   „   *rugosa* Drap.

†§*Conulus fulvus* Drap.

  **Cyclostoma elegans* Müll.

  §*Helix aculeata* Müll.

†§   „   *arbustorum* Linn.

†§   „   „   var. *alpestris* Ziegler

*†§ „   *caperata* Mont.

*†§ „   *concinna* Jeff.

*†§ „   *cricetorum* Müll.

†§   „   *fruticum* Müll.

*†§ „   *hispida* Linn.

  †   „   *lamellata* Jeff.

†§   „   *lapicida* Linn.

*†§ „   *nemoralis* Linn.

  §   „   *obvoluta* Müll.

*†§ „   *pulchella* Müll.

  †   „   *rufescens* Penn.

*†   „   *virgata* Da Costa

†§*Hyalina cellaria* Müll.

†§   „   *crystallina* Müll.

†§   „   *nitida* Müll.

†§   „   *nitidula* Drap.

†§   „   *radiatula* Ald.

  †*Hydrobia marginata* Mich.

INVERTEBRATA (*cont.*).

MOLLUSCA (*cont.*).

Gasteropoda (*cont.*).

†*Limax agrestis* Linn.
† „ *arborum* Bonch. Chant.
† „ *lævis* Müll.
†§*Limnæa auricularia* Linn.
*†§ „ *palustris* Müll.
*†§ „ *peregra* Müll.
†§ „ *stagnalis* Linn.
*†§ „ *truncatula* Müll.
†§*Patula rotundata* Müll.
† „ *ruderata* Stud.
†§ „ *pygmæa* Drap.
†§*Physa hypnorum* Linn.
†§ „ *fontinalis* Linn.
†§*Planorbis carinatus* Müll.
†§ „ *complanatus* Linn.
† „ *corneus* Linn.
†§ „ *contortus* Linn.
† „ *fontanus* Lightf.
†§ „ *glaber* Jeff.
§ „ *nautilus* Linn.
† „ *nitidus* Müll.
*†§ „ *spirorbis* Müll.
†§ „ *vortex* Linn.
*†§*Pupa marginata* Drap.
† „ *muscorum* Linn.
† „ *umbilicata* Drap.
*†§*Succinea elegans* Risso
*†§ „ *putris* Linn. var. *major*
† „ „ var. *minor*
*†§ „ *oblonga* Drap.
*†§*Valvata cristata* Müll.
*†§ „ *piscinalis* Müll.
† „ „ var. *antiqua* Morr.
†§*Vertigo angustior* Jeff.
†§ „ *antivertigo* Drap.

INVERTEBRATA (*cont.*).

MOLLUSCA (*cont.*).

Gasteropoda (*cont.*).

†*Vertigo edentula* Drap.
† „ *minutissima* Hartm.
†§ „ *moulinsiana* Dupuy
† „ *pusilla* Müll.
†§ „ *pygmæa* Drap.
*†§*Zua lubrica* Müll.

Lamellibranchiata.

†*Sphærium (Cyclas) lacustre* Müll.
*†§ „ „ *corneum* Linn.
†§*Cyrena (Corbicula) fluminalis* Müll.
*†§*Pisidium amnicum* Müll.
†§ „ *fontinale* Drap.
*†§ „ „ var. *henslowana* Shepp.
† „ *nitidum* Jenyns
† „ *pulchellum* Jenyns
†§ „ *pusillum* Gmel.
†§*Unio littoralis* Linn.
†§ „ „ var. *limosa* Nils.
†§ „ *pictorum* Linn.
† „ *tumidus* Retz. ?
Anodonta sp.

CRUSTACEA.

§*Cypris reptans* Baird
†*Candona compressa* Koch.
† „ *candida* Müll.

PLANTÆ.

†Spores and stems of *Chara*
†*Salix* (probably *S. repens*)

There are some specially interesting features in the molluscan fauna of this gravel terrace which have been pointed out by Mrs Hughes (*op. cit.*). None of the Mollusca are totally extinct, though seven species (*Cyrena*

fluminalis, Unio littoralis, Unio pictorum var. *limosa,
Hydrobia marginata, Helix fruticum, Paterula ruderata*
and *Clausilia pumila*) no longer exist in Britain, and
some six or eight species are no longer found in the
neighbourhood of Cambridge. "The shell which seems to
indicate the greatest change of conditions is the *Cyrena
(Corbicula) fluminalis.*" It first appeared in Britain in
the Norwich Crag, but at the present day is not found
nearer than Sicily, the Nile and the rivers of Asia Minor
and Syria. Finally, "the whole gravel fauna seems to
point to continental conditions, when Europe was connected with Africa, and England was united with France
and Belgium."

(b)   *The Intermediate Terrace.*   (Cam River system.)

It is almost impossible to distinguish this terrace
in the tributary valleys of the Cam, though there are
many patches of gravel which are on a higher level than
the deposits on the banks of the existing streams but on
a lower level than the gravels above described. There
has been constant re-arrangement as well as erosion of the
older gravels, and they have been incorporated into newer
deposits during the meandering of the streams from side
to side, to which process the widespread deposits of gravel
are generally due. Such is the case in the valleys of the
Granta and Lin above Shelford, and in that of the Bourn.

It is near Shelford that we first get this Intermediate
Terrace plainly shown. From this village a long broad
strip of gravel runs almost due north to Cambridge,
occupying an old channel between the Barnwell Terrace
on the low Trumpington ridge to the west and the Chalk
slopes to the east. Numerous pits around Shelford are

opened in this gravel, which reaches a thickness of 12 ft. and contains fluviatile and land shells (see list).

From Shelford the spread of gravel accompanies the Vicar's Brook to Cambridge, cutting through the older Barnwell Terrace east of Trumpington and broadening out into a wide sheet which underlies the greater part of the town of Cambridge. At the Leys School this gravel lies at the surface, and close to this spot the combined streams of the Rhee and Bourn joined the Cam. Opposite Grantchester on the east bank of the Cam we find gravels of this age. North of Downing College and under the northern part of the town the gravel occurring at the surface may be rather more recent, for its level is on an average 11 ft. lower. A ridge of Gault comes to the surface on Fenner's Ground and runs across Parker's Piece on the eastern side of Christ's Piece to Maids' Causeway, separating the Barnwell from the Intermediate Terrace which is banked against the ridge.

On the west side of the river the same Intermediate Terrace is traceable through Newnham to the gardens of Trinity College. It has an average thickness of 9 or 10 ft. and it lies about 40 ft. above sea level, and is therefore about 10 ft. lower than the Barnwell Terrace.

The extensive pits at Chesterton on the north side of the river are in gravel of the same age, and from them have been obtained many mammalian bones (see list).

From Chesterton the terrace extends to Milton as a sheet of gravel three-quarters of a mile broad, running parallel to the Cam for some distance. Beyond Milton it diverges from the present river and trends away to the north, widening out after passing between Landbeach and Waterbeach. From here northwards a strip of Gault intervenes between this terrace and the recent alluvium of the river. A short distance beyond Denny Abbey the gravel disappears from view, the greater part of it having been removed by denudation and the remainder buried beneath the deposits of the fen.

Everywhere along the valley of the Cam, but especially on the Gault in the neighbourhood of Cambridge, there are troughs with intervening ridges running approximately parallel to the valley and representing old stream beds, but these are now all levelled up, and are detected only by drainage or other similar operations.

LIST OF FOSSILS FROM THE INTERMEDIATE TERRACE.

VERTEBRATA.

MAMMALIA.

> Elephas primigenius Blum. (Chesterton)
> Equus fossilis H. V. Meyer (Chesterton)
> Hippopotamus major Desm. (=H. amphibius) (Chesterton)

INVERTEBRATA.

MOLLUSCA.

> Achatina acicula Müll. (Shelford)
> Helix rufescens Penn. ? (Shelford)
> Planorbis complanatus Linn. ? (Denny Abbey)
> Succinea putris Linn. var. major (Shelford)

(c) The Lowest Terrace. (Cam River system.)

The Lowest Terrace of the Cam River system is only slightly raised above the level of the modern alluvium, and borders the streams in their present courses for long distances. From Chesterford down to Shelford the Granta has an almost continuous fringe of these gravels on both its banks, and they are well shown in the pits east of the line by Whittlesford Station.

In the valley of the Lin from Bartlow down to Abington patches of gravel belonging to this terrace occur at various spots. Near Bartlow Station in the 10 ft. of gravel and loam tusks of *Elephas primigenius* and shells of *Helix, Bithinia, Pupa, etc.* have been found. From Abington a continuous strip of gravel extends to Shelford. At Shelford we find that the river formed by the union of the Lin and the Granta turned westward to Hauxton instead of flowing northward as at the time of the formation of the Intermediate Terrace.

Near Hauxton the river joined the Rhee, but owing to the opposing barrier of Chalk Marl which here stretched across its course denudation was delayed at this point, and above it the river meandered about, cutting into the sides of its valley and giving rise to wide alluvial flats liable to frequent floods. Finally the river cut its channel through the barrier deep enough to carry off the water freely. This is the probable explanation of the large spread of gravel near Hauxton, sections of which may be seen in the pits from which the Chalk Marl is now extracted for cement. Some geologists suppose that a lake existed behind the barrier[1].

The Bourn River, which flows into the united Cam and Rhee south of Grantchester, shows strips of gravel along its course from Bourn eastwards to Grantchester which may belong to this terrace.

The road along the "Backs" runs at the foot of the Intermediate Terrace, which is marked by the slightly higher ground on which lie the Fellows' Gardens of King's, Clare, and Trinity Colleges. The flat expanse of ground of Chesterton Common and the "De Freville Estate" are situated on the Lowest Terrace. Further

[1] *Mem. Geol. Surv. Explan. Quart. Sheet* 51, S.W. p. 105.

north a narrow strip runs by the side of the railway line to Ely, and another strip by Milton Fen is banked against the higher terrace. A patch by Horningsea seems to indicate the former existence of a small tributary from the south, perhaps the same that deposited the Wilbraham gravels (see below). Beyond Waterbeach the gravel is found below a thin alluvial covering and probably extends north beneath the fen between Denny Abbey and Upware. Beyond this point this terrace has not been positively identified.

List of Fossils from the Lowest Terrace.

VERTEBRATA.

MAMMALIA. *Bos longifrons* Owen
Elephas primigenius Blum. (near Chesterton)

INVERTEBRATA.

MOLLUSCA. *Bithinia tentaculata* Linn.
Helix sp.
Pupa marginata Drap. ?
Succinea putris Linn. var. *major*

Gravels near Wilbraham and Fulbourn.

A large lake appears to have existed between Wilbraham and Fulbourn at some period subsequent to the institution of the present river-system, and we find its thin deposits of gravel and sand spreading over the low ground now occupied by Wilbraham Fen. In fact this sheet of water has only been drained during the last century, but it must have stood at a higher level before the deepening of its outlet near Quy-cum-Stow. Though this lake must have been of great antiquity, yet we find patches of gravel on the higher ground below the level

of the ancient river-system series but above the deposits
of the lake, which apparently were accumulated before
the lake existed, and mark out the course of a feeder of
the Cam. Such patches are found at Little Wilbraham,
near Quy Mill, and N.E. of Horningsea.

Gravels of the valley of the Ouse.

There are two terraces more or less distinct in the
upper part of the valley of the Ouse. They are raised
a few feet above the level of the fen and of the recent
river-deposits and are generally separated from them
by a strip of Jurassic clay.

The upper terrace appears to correspond to the
Barnwell Terrace. The large patch of gravel, sand and
loam between Fenny Stanton and Fenny Drayton belongs
to it, and so does another patch between Holywell and
Needingworth on the north bank of the river. Between
Bluntisham, Colne, and Somersham are some more strips
and patches of this gravel, which have here yielded *Bos
primigenius* and some shells including *Cyrena fluminalis*
and *Cardium edule*. The large spread of gravel north of
Somersham also belongs to this terrace, and thus it
appears probable that at this period the Ouse turned off
northward near Bluntisham to flow into the Wash.

The lower terrace of the Ouse is found around
St Ives, and the town stands on a broad portion of it. At
Swavesey and on the west and north side of Bluntisham
Fen smaller patches occur, and at "The Holcrofts" a
larger expanse of similar gravel is found.

Between this point and the Cam there is a complete
gap, and not until Ely is reached do we get any gravel
which can be assigned to either stage. Here between

Roslyn Hill and the town of Ely an isolated patch of gravel extends which may belong to the lower terrace; in it have been found *Bos longifrons, Cervus,* and *Equus fossilis*[1]. At Littleport on the high ground a mass of gravel is mapped, but to what terrace it belongs is uncertain. The old course of the Ouse (Old Welney River) is marked at Butcher's Hill by sand and gravel probably belonging to a lower terrace.

Gravels of the tributaries of the Ouse.

In the valley of the Kennett there are large spreads of gravel. The different terraces have not been made out, but the mammalian bones found in a pit in the gravel near Kentford Church[2] show that probably some of the gravel belongs to the age of the Barnwell Terrace (see p. 238). In Chippenham Field, more than a mile N.N.W. of the church, gravel is said to have been worked to a depth of 60 feet before Chalk was touched[3]. There are broad patches of gravel between Fordham and Worlington, but their age and significance are doubtful.

The wide patches of gravel on the southern side of the valley of the Lark nearly join those of the Kennett, but they lie in the county of Suffolk.

[1] *Geol. of Fenland*, p. 183, and *Mem. Geol. Surv. Explan. Quart. Sheet* 51, N.E. p. 81.

[2] *Mem. Geol. Surv. Explan. Quart. Sheet* 51, N.E. p. 76.

[3] *ibid.* p. 77.

THE FEN DEPOSITS.

The whole of the northern part of Cambridgeshire is occupied by fenland[1]. The southern boundary of this tract runs, roughly speaking, from Over in the west through Willingham and Cottenham to a little north of Denny Abbey. Here it turns sharply southwards to Waterbeach, crosses the river Cam, and runs round the southern margin of Bottisham Fen, sending promontories down towards Horningsea and Swaffham Bulbeck. Thence it sweeps round the east side of Burwell Fen, stretches out to Fordham and runs in a wavy line past Wicken Fen and round Soham Mere towards Mildenhall.

The superficial deposits of the Fenland consist of gravel, peat and silt. It was formerly held that the beds could be classified into a lower stratum of peat, a middle bed called the "Buttery Clay" and an overlying bed of peat termed "the Upper Peat." But according to the recent work of the Geological Survey[2] no such classification holds good for the whole area.

Gravel forms the basement bed of all the fen deposits and with but few exceptions appears to exist as a continuous and persistent sheet beneath them[3]. Above this comes a variable series of beds of peat and silt; in some places two beds of peat are shown separated by silt; in others peat represents the whole fen series down to the gravels; in others gravels and silts preponderate; and in

[1] For a full description of the whole Fenland see Miller and Skertchly, *The Fenland* (1878) and *Mem. Geol. Surv. Geology of the Fenland* (1877).

[2] *Mem. Geol. Surv. Explan. Sheet* 65 (1886), pp. 118—119.

[3] *Mem. Geol. Surv. Geol. of Fenland* (1877), p. 183.

others there is no peat at all. There are many instances
of the rapid alternation and variation of beds within short
distances.

Speaking generally, the northern part of the Fenland
may be said to consist of marine silt and the southern
part of peat. The southern limit of the silt marks the
shore of the great bay—the predecessor of the Wash:
but owing to the frequent invasions of the sea into
the flat peat-country this line is complicated and indefinite.

But this great silt-covered plain has been advancing
northwards from prehistoric to Roman and mediæval
times, and the process is going on still at the present
day. So long as the climatic conditions were favourable
the peat continually encroached on it from the south,
but as the climate became drier the growth of the peat
diminished in vigour and finally ceased. Nevertheless
the sea has continued to retreat and the process of the
silting up of the estuaries of the Ouse and of the other
rivers flowing into the shallow bay has gone on incessantly.

The ancient estuary of the river Ouse lies partly in
Cambridgeshire. The course of the river above Littleport
was originally to the north-west, and it passed up what
are now called the Old Croft and Old Welney Rivers by
Welney to Upwell and thence through Wisbech to the
sea. Sections through the beds along this line show an
alternating series of loams, clays and peat-seams with
estuarine and dwarfed marine shells[1]. Before the end of
the thirteenth century[2] this river Ouse—the Great Ouse—
was artificially diverted into the Little Ouse by a cross
cut to Brandon Creek. The Welney river then became

---

[1] *Mem. Geol. Surv. Explan. Sheet* 65, pp. 126—131.
[2] *Mem. Geol. Surv. Geol. of Fenland*, pp. 86—90.

a branch of the Nene, till the final draining of the fens closed its course for ever. The long prolongation of the tract of silt from Wisbech south-eastwards to Littleport and the ditch called the Croft River mark the course of this once important but now vanished river[1].

The aspect of the Fenland is well known, but the physical features of the tract occupied by peat are not identical with those of the broad northern fringe of silt-land. The peat-land forms a perfectly flat open plain, quite treeless except for the lines of poplars and willows along the water courses, and devoid of hedgerows and villages. The black colour and the peculiar nature of the soil are also its special characteristics. The silt-land has on the other hand a somewhat uneven and undulating surface and is dotted with trees and villages.

(a) The Peat.

The peat of the Fenland is true marsh- or bog-peat and consists of the accumulated remains of mosses and swamp-loving plants. Rushes (*Juncus*), reeds (*Arundo*), sedges (*Carex*), mosses (*Hypnum*), stoneworts (*Chara*), bladderworts (*Utricularia*), confervæ, and other aquatic plants, together with the roots of willows and of other trees, and the wood of the buried forests of oak, elm, fir, birch, yew, alder, and willows, all enter into its composition.

During the greater part of the period which the peat represents a climate must have prevailed favourable to its growth, but occasionally drier intervals intervened when the surface of the peat became firm and the waters retreated and forests of large trees grew up. When the area was again flooded the trees were destroyed.

[1] *Mem. Geol. Surv. Geol. of Fenland*, Plates i. and xv. and p. 140.

At some date, probably within historic times, the peat ceased to be formed owing to the gradually increasing dryness of the climate due to embankments, drainage, and cultivation.

The upper part of the peat is always weathered to a crumbling black or dark brown material, but beneath this surface layer it is fibrous, soft, and elastic, with distinct plant remains (rushes, reeds, etc.) and affords the best fuel. The lowest part is almost entirely composed of moss (*Hypnum*) and dries to a golden yellow colour. All the peat beneath the weathered layer shows bedding more or less distinctly; the lower beds are compressed and more compact owing to the weight of the overlying layers.

In addition to the vegetable remains in the peat, there are present the elytra of beetles, the bones of mammals, human implements of Neolithic type, and objects of every later age which have accidentally been entombed or have sunk into it.

The peat occurs in beds of more or less persistence in thickness and extent, but the division into an Upper and a Lower Peat, as has been stated, does not hold good for the whole area. Interbedded layers of silt and clay occur at various levels and point to interruptions in the vegetable growth, and to river floods and incursions of the sea. Sometimes the alternations of marine and freshwater conditions must have been very rapid.

The thickness of the peat naturally varies, because in some localities there was no continuous growth of vegetable matter, and here and there it is split up or replaced by marine silts and clays. In other spots we get an uninterrupted mass of peat. It should also be remembered that in those places where long and excessive drainage has been carried on, the peat has shrunk considerably. Thus at Hilgay Fen this shrinkage amounted to 52 inches in 26 years[1].

The peat on the whole appears to be thicker in the south than in the north, as was to be expected owing to its comparative freedom in the south from marine irruptions. A thickness of 18 feet has been measured in the parish of Earith, but the average thickness is much less.

Where beds of silt, etc. are intercalated the individual peat beds measure only from a few inches to a foot and a half in thickness; and the contemporaneous denudation and breaking up of the peat is often apparent.

Along the margins of the peat-land the peat thins out, and is often hard to separate from the modern moory turf.

The "buried forests" in the peat are found at various levels, pointing to the repeated occurrence of conditions favourable to the growth of trees during the period in which the peat was formed. The trees of which the forests consist are oak, elm, fir, birch, yew, alder, and several species of willow. The timber is stained black or grey through the action, of the peat, but the wood of the oak is sometimes fairly sound. Some of the trees must have been of enormous size; a trunk of an oak 70 feet long was recently discovered in Bottisham Fen, and bigger ones have been recorded. But the majority of the trees were not exceptionally large. In the Isle of Ely a succession of five forests has been determined. The lowest consists of oaks and yews, the oaks being found where the subsoil is of clay, and the yews where it is of sand. The next forest consists of the same kinds of trees in peat. The two succeeding forests are of firs (*Pinus sylvestris*), while the highest and latest contains

[1] *Mem. Geol. Surv. Explan. Sheet* 65, p. 123.

the stumps, trunks, twigs, and roots of sallows, willows, and alders[1]. Most of the trees are mere stumps broken off two or three feet above the roots, but the stumps with the roots attached are often present. The prevailing direction of the fallen trees in the peat is from S.W. to N.E. and it is noticeable that the trees now growing along the dykes and streams incline also towards the N.E. owing to the prevalence of south-westerly winds. In prehistoric and early historic times we may therefore conclude that the same winds prevailed. This mode of occurrence, taken in conjunction with the facts that the peat was for the most part formed earlier than Roman times and that the tree-trunks have plainly been broken off and not hewn down, renders untenable the view that the forests were cut down by the Romans. What apparently was the cause of the destruction of the forests was the flooding of the area owing to the rivers overflowing their banks and converting the moderately dry forest-land into expanses of stagnant water. The growth of the trees would at first be checked, and ultimately they would die and be blown down by the south-westerly gales. In course of time they would be covered by the soft but preservative mass of decaying vegetation accumulating around their stems and roots[2].

Palæontology of the Peat. The fossils of the peat are mostly the remains of plants and vertebrates, and the state of preservation of the bones is very characteristic. They are stained a deep chocolate brown or black, and are usually in a perfect condition. Frequently the bones of a whole skeleton are found associated together, showing that there has been no drifting or washing about of

---

[1] *Mem. Geol. Surv. Explan. Sheet* 51, N.E. pp. 93—97.

[2] *Mem. Geol. Surv. Explan. Sheet* 65, p. 126.

the carcase, but that the animal was mired and sank into the soft yielding swamp, or died quietly in the swamps and was buried in course of time beneath the dense vegetable growth.

The Mammalia from the peat partly belong to species which are everywhere extinct, partly to species which are no longer found in England, and partly to species which still inhabit our country. The remains of *Rhinoceros* and *Hippopotamus* which have been recorded from the peat, were long ago shown to be derived from older deposits, and the statement that the Irish Elk (*Cervus giganteus*) and the Reindeer (*Cervus tarandus*) have been found in the peat of this area still requires proof.

In the upper part of the peat human implements of Neolithic type are not rare, and the skull of a *Bos primigenius* with a flint weapon imbedded in its forehead was found in 1863 in Burwell Fen[1] and is now in the Woodwardian Museum. Bronze and iron implements and other objects belonging to later ages are also found in the upper layers.

FOSSILS OF THE PEAT[2].

PLANTÆ.

Alnus glutinosa
Betula alba Linn.
„ *nana* Linn.
Chara sp.
Fagus sylvaticus Linn.
Fraxinus sp.
Hypnum fluitans Dill.
„ *filicinum* Vill.
Juncus obtusiflorus Ehrh.

[1] *Proc. Camb. Antiq. Soc.* ii. p. 285. *Geol. Mag.* Dec. 2, vol. i. (1874), p. 492. Bonney, *Camb. Geol.* p. 59.
[2] *Mem. Geol. Surv. Geol. of the Fenland* (1877), App. K, p. 320.

PLANTÆ (*cont.*).

> *Lastrœa* sp.
> *Pinus sylvestris* Linn.
> *Quercus robur* Linn.
> *Salix caprea* Linn.
>   „   *cinerea* Linn.
>   „   *repens* Linn.
> *Taxus baccatus* Linn.
> *Ulmus* sp.
> *Utricularia* sp.

INVERTEBRATA.

INSECTA.

> *Copris lunaris* Linn.
> *Donacia simplex* Fabr.
> *Elater* sp.
> Various indeterminable *Neuroptera*

VERTEBRATA.

PISCES.

> *Esox lucius* Linn.

REPTILIA.

> *Emys orbicularis* Linn. (extinct in England)

AVES.

> *Botaurus stellaris* Linn.
> *Cygnus musicus* Bechs.
>   „   *olor* Gmelin.
> *Fulica atra* Linn.
> *Pelecanus onocrotalus* Linn. ?
> *Podiceps cristatus* Linn.
> *Querquedula crecca* Linn.

MAMMALIA.

> *Bos longifrons* Owen (extinct everywhere)
>   „   *primigenius* Boj.   (ditto)
> *Canis lupus* Linn. (extinct in England)
>   „   *vulpes* Linn. (existing in England)
> *Castor fiber* Linn. (extinct in England)
> *Cervus capreolus* Linn. (existing in England)
>   „   *elaphus* Linn.   (ditto)

VERTEBRATA (*cont.*).

    MAMMALIA (*cont.*).

        Lutra vulgaris Erxl. (existing in England)

        Martes abietum Gmelin   (ditto)

        Sus scrofa Linn. (extinct in England)

        Ursus arctos Linn.   (ditto)

(b) *The Clays and Silts.* (= 'Fen Silt' of Survey Memoir.)

Between the layers of peat there are intercalated in places thin beds of clay and laminated silt, indicating alternations of marine and freshwater conditions as already mentioned. These intercalations are particularly frequent on the borderland between the peat-land and the silt-land where there were frequent floodings by the sea of the marshy tracts[1].

In the neighbourhood of Cambridge there is a fairly persistent bed of clay which is called the "Buttery Clay," and it separates the peat into an upper and a lower division[2], but apparently this is only a local character, for the "Buttery Clay" merges into a clay containing *Scrobicularia piperata* and therefrom called the "Scrobicularia Clay," and this also passes into the silty or sandy deposits called "warp." In fact these three deposits —"Buttery Clay," "Scrobicularia Clay" and "Warp"— are merely local facies of one and the same deposit[3]. The clay, according to Prof. Bonney, is usually about 6 feet thick but has been known to measure as much as 30 feet. In colour it is usually light or dark blue and purple, but is often mottled. Sometimes it contains carbonaceous

---

[1] See Sections in *Mem. Geol. Surv. Geol. of Fenland* (1877), pp. 146—151.

[2] Bonney, *Camb. Geol.* p. 57.

[3] *Geol. of Fenland*, pp. 173, 174. *Mem. Geol. Surv. Sheet* 65, p. 133.

matter. "When wood is plentiful the bright blue phosphate of iron (Vivianite) occurs in amorphous earthy lumps and streaks, varying in size from fine specks to bits as large as a bean." The clay is rarely bedded, is of a soft 'buttery' or unctuous character, very rarely contains stones, and is fairly free from sand.

The commonest and most characteristic fossil is *Scrobicularia piperata*, and this shell is often found with both valves adherent in the position in which it lived. The bones of whale, seal, walrus, and grampus have also been found in it, with those of *Bos* and *Sus*. Foraminifera are very plentiful.

The whole lithological character of the clay and its included fossils prove that it is a tidal deposit; and the process of accumulation of similar materials is now going on in the Wash. The finer argillaceous material corresponding exactly to the "buttery clay" is thrown down in rather deeper water than the sandy silt which is the modern counterpart of the 'warp.'

The 'warp' is a very fine sandy deposit, finely laminated and containing a great abundance of Foraminifera, shell fragments, and minute flakes of mica. Beds of 'warp' and clay alternate abruptly, with thicknesses varying according to locality. A gradual transition from the clay into the sandy 'warp' can occasionally be traced in a horizontal direction, as for instance near Croyland[1].

MOLLUSCA FROM THE 'BUTTERY CLAY.'

Cardium edule Linn.
Mytilus edulis Linn.
Ostrea edulis Linn.
Scrobicularia piperata Linn.
Tellina balthica Linn.
Rissoa sp.

[1] *Geol. of Fenland*, p. 178.

From the Fen Silt, as a whole, including the clay and the warp, the following list is given in the Survey Memoir on the Fenland :—

FORAMINIFERA (very plentiful).

MOLLUSCA.

    Cardium edule Linn.
    Mytilus edulis Linn.
    Ostrea edulis Linn.
    Pisidium amnicum Müll.
    Scrobicularia piperata Linn.

Gasteropoda.

    Bithinia tentaculata Linn.
    Helix pulchella Müll.
    Hydrobia (Rissoa) ulvæ Pen.
    Limnæa peregra Müll.
       „   *stagnalis* Linn.
    Planorbis carinatus Müll.
       „   *complanatus* Linn.
       „   *lævis* Alder.
       „   *vortex* Linn.
    Physa fontinalis Linn.
    Valvata cristata Müll.

MAMMALIA.

    Balæna mysticetus Linn.
    Delphinus turtio Fabr.
    Orca gladiator Gray
    Phoca vitulina Linn.
    Phocæna crassidens Owen
    Trichechus rosmarus Linn.

(c) *The Shell Marl.*

In the eastern part of the peat-land there is found a layer of white friable marl, full of freshwater shells, a foot or two below the surface of the peat. It is met with in the tract of country extending from Burwell Fen to

Stretham Mere, Burnt Fen, Soham Mere, Sedge Fen and Lakenheath[1], and again between Littleport and Downham[2], not as a continuous sheet but in patches. It varies in thickness from about 1 to 5 ft.

It contains abundant remains of *Chara*, and the formation of it is to be attributed to the decay of this plant in shallow meres. It is well known that the stems of the common species of *Chara* are encrusted with carbonate of lime, and it has been noticed that they now grow in dense masses to the exclusion of other aquatic plants and when they die leave a deposit of carbonate of lime at the bottom of the water.

For the formation of this shell-marl we must conclude[3] that the water in the shallow meres was free from mud, and was charged with a considerable amount of carbonate of lime, the water being derived from the Chalk. This was unfavourable to the growth of peat, though not to the existence of molluscs, fish, and some aquatic plants.

The following Mollusca and Ostracoda have been recorded from the Shell-Marl of South Level (= L), from the alluvium of Soham Mere (= M), from the shell-marl along the railway between Littleport and Downham Market (= R)[4], and from "an old lacustrine deposit, overlaid by 5 or 6 feet of marl," at Whittlesea (= W)[5].

MOLLUSCA.
  Gasteropoda.
    Bulla ? L.
    Bithinia tentaculata Linn. L. M. R.

[1] *Mem. Geol. Surv. Explan. Sheet* 51, N.E. p. 98.
[2] *Q. J. G. S.* vol. VI. (1850), pp. 451—453.
[3] *Mem. Geol. Surv. Explan. Sheet* 51, N.E. p. 99.
[4] *Ibid.* pp. 101—102, and *Explan. Sheet* 65, p. 140.
[5] G. S. Brady, Crosskey and Robertson, *Mon. Post-Tert. Entomostr.* (*Palæont. Soc.* 1874), p. 108.

MOLLUSCA (*cont.*).
 Gasteropoda (*cont.*).
  Helix pulchella Müll. R.
  Limnæa auricularia Linn. L. M.
   ,, *palustris* Müll. M.
   ,, *peregra* Müll. L.
   ,,   ,, 2 vars. R.
   ,, *stagnalis* Linn. L. R.
  Paludina contecta Millett
  Physa fontinalis Linn. R.
  Planorbis carinatus Müll. M.
   ,, *complanatus* Linn. L. M.
   ,, *corneus* Linn. M.
   ,, *lævis* Alder R.
   ,, *marginatus* Drap. R.
   ,, *spirorbis* Linn. L.
   ,, *vortex* Linn. R.
  Succinea elegans Risso. M,
   ,, *putris* Linn. L.
  Valvata cristata Müll. M. R.
   ,, *piscinalis* Müll. L. M.

 Lamellibranchiata.

  Pisidium amnicum Müll. L. M.
   ,, *nitidum* Jenyns M.
   ,, *obtusale* Lam. (? var. of *P. pusillum*) R.
   ,, *pusillum* Gmel. R.
  Sphærium corneum Linn. L. M.

CRUSTACEA.
 Ostracoda.
  Candona albicans Brady W.
   ,, *candida* Müll. R. W.
   ,, *compressa* Koch W.
   ,, *lactea* Baird W.
   ,, *pubescens* Koch R.
  Cypridopsis Newtoni B. and R. W.
  Cypris incongruens Ramd. R.
   ,, *lævis* Müll. R. W.
   ,, *ovum* Jur. W.

CRUSTACEA (*cont.*).
  Ostracoda (*cont.*).
    Darwinella Stevensoni B. and R. W.
    Herpetocypris (*Candona*) *reptans* Baird R. W.
    Ilyocypris (*Cypris*) *gibba* Ramd. R. W.

Economics of the Fen deposits. The use of peat
for fuel is general in the Fen districts and it is dug for
this purpose in many places in the neighbourhood of
Cambridge and carried by barge and cart for miles round.
On Burwell Fen and again at Coveney west of Ely
the process may be observed. Long parallel trenches
about a yard in width and five or six yards apart are cut
in the peat; the upper weathered layer is cut away
first in rough lumps called 'hods'; the underlying un-
weathered portion is cut into brick-like blocks called
'cesses' or 'turves' with a wooden ironshod spade called
a 'becket[1].'

The clay mixed with sand is used to make an inferior
kind of brick. The shell-marl is dug in Sedge Fen for
manuring the land.

Agriculturally the Fenland forms one of the richest
districts in the kingdom, the water of the fens being
now under the complete control of an artificial system of
drainage[2].

[1] *Mem. Geol. Surv. Geol. of Fenland*, pp. 135—139.
[2] Miller and Skertchly, *The Fenland*, pp. 562—566.

RECENT ALLUVIUM, ETC.

Bordering the present rivers and their tributaries are strips of flat marshy land of variable width composed of alluvial deposits of sandy loam and mud of recent origin and comparatively slight geological importance.

Black peaty earth with land and freshwater shells is often found immediately beneath the surface soil on the low level land, marking the site of old marshes and swamps, many of which have only been drained within the last century or so.

Blown sand derived from the weathering of the sandy Boulder Clay covers large tracts to the north-east of Newmarket around Mildenhall, and pebbles polished by the attrition of the grains of sand are found in that neighbourhood[1]. Rain-wash occurs here and there on the slopes and at the foot of the Chalk hills.

'Warp' and 'trail' are terms which have been employed to denote the two kinds of subsoil found in this area.

Beneath the vegetable mould lie the soil and subsoil, which are composed partly of the débris of the subjacent undisturbed rock and partly of material from a distance. The upper part of this surface covering has been called the 'warp,' to be distinguished from the tidal warp of the Fens, and the lower portion, which occurs in furrows

[1] *Mem. Geol. Surv. Explan. Sheet* 51, N.E. p. 88.

and troughs, has been called the 'trail[1].' The latter is devoid of organic remains and consists usually of a marly clay with stones and patches of gravel. It shows evidence of having been subjected to considerable pressure, and contains foreign material. The so-called 'warp' rests on the trail unconformably with signs of erosion, and is composed of dark grey soil. It derives its materials from the 'trail' and from the neighbouring solid rocks, and is the result of the growth of vegetation and of the operation of subaerial agencies. It contains land-shells, some of which are either no longer living or are very rare in the district.

[1] Rev. O. Fisher, *Q. J. G. S.* xxii. (1866), p. 562. *Mem. Geol. Surv. Explan. Sheet* 51, S.W. p. 113.

THE ANTIQUITY OF MAN IN THE DISTRICT.

In this district there are but few traces of the presence of Palæolithic man, though in the adjoining counties
river-drift implements have been found in considerable
numbers[1]. A few such implements have been recorded
from the gravels of the Cam valley. Prof. Seeley[2] discovered a rib bone of an elephant, marked with peculiar
sharp cuts, in a marl bed at Barnwell associated with the
characteristic shells of the Barnwell Terrace of river-gravel,
but whether the cuts on this bone were made by man or
in some other way is a disputed point. A flint hache,
closely corresponding to some found at Biddenham in
Bedfordshire in gravels of the age of the Barnwell
Terrace, is stated to have been found in the gravel of
the Barnwell pit and was described by Mr A. F. Griffith
in 1878[3].

Other worked flints of Palæolithic types have been
picked up on heaps of gravel near Cambridge[4], and one
is stated to have been found in the Barrington gravels.

[1] Sir J. Evans, *Ancient Stone Implements, etc. of Great Britain*, 1872,
ch. xxiii.

[2] *Q. J. G. S.* vol. XXII. (1866), p. 475.

[3] *Geol. Mag.* Dec. 2, vol. v. (1878), p. 400.

[4] Sir J. Evans, *op. cit.* p. 485.

Prof. Bonney[1] mentions a flint flake from the gravel of Midsummer Common, and a rude implement believed to have come from the Observatory gravel. The pointed implement from the neighbourhood of Burwell, which was described by Prof. Babington[2], is of doubtful genuineness and probably is a forgery[3]. An undoubted palæolithic implement has been found in the March gravels[4], and it has been suggested[5] that the implement-bearing gravels of Brandon in Suffolk are allied to those of the ancient river-system in our county to which the March gravels are ascribed.

Near Kennet Mr A. F. Wright[6] found in 1886 several palæolithic implements on the surface of a field, and in a ballast pit near Kentford Church a number of implements and flakes were discovered in the upper layers of gravel associated with mammalian remains (*Elephas primigenius, Hippopotamus, Rhinoceros, Bos, Cervus*)[7].

The scarcity of implements around Cambridge is remarkable and difficult to explain, for in the neighbouring valleys they occur in considerable abundance[8]. Thus in the valley of the Great Ouse around Bedford, in the valley of the Lark near Bury St Edmunds, Icklingham and Mildenhall, and in the valley of the Little Ouse at Thetford in Norfolk, Santon Downham and Brandon the large dis-

[1] Bonney, *Camb. Geol.* p. 55.

[2] *Antiq. Comm. (Camb. Antiq. Soc.)*, vol. II. p. 201.

[3] Sir J. Evans, *op. cit.* p. 485.

[4] *Mem. Geol. Surv. Explan. Sheet* 65, p. 113.

[5] *Ibid.* pp. 71, 72.

[6] A. F. Wright, *Nature*, vol. XXXIV. (1886), p. 521.

[7] *Mem. Geol. Surv. Explan. Quart. Sheet* 51, N.E. p. 76.

[8] Sir J. Evans, *op. cit.* pp. 479, 486, 494. *Mem. Geol. Surv. Explan. Quart. Sheet* 51, N.E. pp. 73—79.

coveries of Palæolithic implements in the river-gravels
are famous. In places they occur in such numbers that
Mr H. Prigg[1] declared that within the watershed of the
Lark and Little Ouse over 6000 had been found within
the last twenty years. Frequently they are associated
with the fauna of the Barnwell Terrace, as, for example,
around Bedford.

Of the presence of Neolithic man we have abundant
evidence, and examples of his handiwork have been found
in considerable numbers. Sir John Evans mentions
(*op. cit.*) rough-hewn celts from Wicken and Bottisham
Fens, Burwell, and Bartlow Hills, polished greenstone
celts from Coton, Manea, Burwell and Bottisham, stemmed
and barbed arrow-heads from Aldreth and Burwell Fen,
perforated hammer-heads from Reach and Newmarket,
perforated axes from Ely and Chatteris Fen, and many
others. Around Burwell implements and weapons appear
especially common, and the same author records adzes,
gouges, whetstones, knives, daggers, mealing stones, etc.
in addition to those above mentioned from that locality.
Jet ornaments have been found in Soham Fen. The
skull of the *Bos primigenius* in the forehead of which
a celt was found imbedded has been previously mentioned
(p. 227). We find that Neolithic man was thus contemporaneous with a fauna of which many members
have now either locally disappeared, or are no longer
living in England, or have become everywhere extinct[2]
(p. 228).

Though no indisputable traces of lake dwellings have

[1] *Proc. Norw. Geol. Soc.* part vi. (1882), pp. 163, 165.
[2] See Boyd Dawkins, *Early Man in Britain.* Skertchly, *The Fenland*,
pp. 339—354.

been found in Cambridgeshire[1], yet in the neighbouring counties at Crowland, Wretham Mere near Thetford, and Barton Mere near Bury St Edmunds undoubted examples have been described[2]. Relics of the Bronze Age[3] are rare, but those of later times are fairly common in the peat of the fens and in the superficial deposits of the district. Such antiquities, however, do not come within the scope of this work.

WATER SUPPLY.

The water supply from the Jurassic clays is comparatively small and is derived from the bands of limestone, nodules and septaria which occur in them. But there is some uncertainty about obtaining water from them, for in wells at Conington and Bluntisham a depth of 300 feet was reached without success, whereas at Redhill, a farm west of Conington, a good spring was met with at a depth of 12 feet. If the Oxford Clay was pierced we might reasonably expect an abundant supply of water[4].

The Lower Greensand is the main water-bearing stratum of the district, and borings into it through the

[1] Mem. Geol. Surv. Geol. of Fenland, p. 246; R. Munro, Lake Dwellings of Europe (1890), p. 459.

[2] Mem. Geol. Surv. Geol. of Fenland, p. 248; Sir C. Bunbury, Q. J. G. S. vol. XII. (1856), p. 355; R. Munro, Lake Dwellings of Europe, p. 455.

[3] Sir J. Evans, The Ancient Bronze Implements of Great Britain (1881); Miller and Skertchly, The Fenland, p. 462.

[4] Mem. Geol. Surv. Explan. Quart. Sheet 51, S.W. p. 128.

overlying beds yield an excellent and constant supply. At the Cambridge Waterworks at Cherry Hinton the following is the section of the shaft to the Gault and the boring. The water rises here to within a few feet of the surface.

		Feet
Chalk Marl 48 feet	Soil and light-coloured marl	6
	Darker clunch	23
	Light-coloured clunch or marl	7
	Greyish chalk	4
	Blue clunch with nodule bed at bottom ...	8
Gault	Slate-grey clay, with a band of small nodules 33 feet from the bottom	125
Lower Greensand	Brown clayey sand, with ferruginous phosphate nodules at bottom, and a hard rock below	2
	Soft brown sand with water	1

In the town of Cambridge the well-borings have reached the Greensand at depths from the surface of 125—200 feet. It is what is called 'hard' water, containing carbonate and sulphate of lime, the solid contents of one gallon varying from 14 to 20 grains[1].

The so-called "fossil springs" from the Cambridge Greensand where the Gault is overlaid by the Chalk Marl are not of much importance (see p. 125).

The Chalk is a most valuable water-bearing formation, and the Burwell Rock especially throws out strong springs (see pp. 128, 129).

The line of saturation on the Chalk hills rises to a considerable height, so that at Heath Farm, Shelford, 156 feet above Ordnance datum, the water stands only 96 feet below the surface, and at Stapleford Windmill, 120 feet above Ordnance datum, it stands 53 feet from

[1] Prestwich. *The Water-bearing Strata of London* (1851), p. 168.

the surface. Under the highest part of the Gog Magog Hills the water-level dips down, probably on account of the line of saturation being lowered by the synclinal in the strata and by the springs which are thrown out by the Burwell Rock where the trough comes to the surface of the ground[1].

For local details and well-sections the Geological Survey Memoirs must be consulted.

The Boulder Clay is sometimes slightly permeable, and wells sunk into it fill slowly with soakage water[2].

The gravels furnish an abundant supply of water where they rest on clay. But the amount of water in the wells is liable to be affected by drought, and it is frequently polluted by sewage or other causes.

The thermal springs[3] near Chatteris were originally believed to owe their temperature (66° F.—74° F.) to deep-seated agencies and to well up along a line of fault, but doubt has been thrown upon this explanation. The wells in which this warm water is found are all very shallow and do not descend to a greater depth than 10—14 feet, only penetrating the fenland deposits and gravel. It has been suggested that some local chemical decomposition may produce the heat.

[1] *Mem. Geol. Surv. Explan. Quart. Sheet* 51, S.W. p. 129, Pl. 5.
[2] *Ibid.* p. 130.
[3] *Mem. Geol. Surv. Geol. of Fenland*, p. 243; Rev. O. Fisher, *Geol. Mag.* vol. VIII. (1871), p. 42; F. W. Harmer, *Rep. Brit. Assoc. Trans. of Sections* (1871), p. 74; *Geol. Mag.* vol. VIII. (1871), p. 143.

APPENDIX.

A LIST of works on the geology of Cambridgeshire was given by Mr W. Whitaker in 1873 to Prof. Hughes, and was published by him for the use of geological students. It was afterwards, in 1881, printed with additions as an Appendix to the Survey Memoir on *The Geology of the Neighbourhood of Cambridge* (Explanation of Quarter Sheet 51 S.W. with part of 51 N.W.).

PUBLICATIONS OF THE GEOLOGICAL SURVEY.

MAPS. (One inch.)

1864. Sheet 52 N.E. (small part). By H. H. Howell.

1864. Sheet 52 S.E. (part).

1869. Sheet 46 N.E. (small part). By W. Whitaker and F. J. Bennett.

1872. Sheet 64 (part). By Prof. J. W. Judd.

1881. Sheet 47 (part). By. W. H. Penning.

1881. Sheet 51 S.W. By W. H. Penning and A. J. Jukes Browne.

1881. Sheet 51 S.W. Drift edition.

1882. Sheet 51 S.E. By F. J. Bennett and J. H. Blake.

1882. Sheet 51 S.E. Drift edition.

1882. Sheet 51 N.W. By W. H. Penning, A. J. Jukes Browne, and others.

1882. Sheet 51 N.W. Drift edition.

1883. Sheet 51 N.E. By S. J. B. Skertchly, F. J. Bennett, and others.

1883. Sheet 51 N.E. Drift edition.

1884. Sheet 47. Drift edition. By W. Whitaker, W. H. Penning, and others.

1886. Sheet 65. Drift edition. By W. Whitaker, H. B. Woodward, S. B. J. Skertchly, F. J. Bennett, A. J. Jukes Browne, and others.

INDEX MAPS. (Four miles to the inch.)

1892. Sheet 12 (includes the southern part of Cambridgeshire).

1897. Sheet 9 (includes town of Cambridge and northern part of the county).

HORIZONTAL SECTIONS.

1882. No. 126.

MEMOIRS.

1856. Decade v. (Plate 5, Fossils from the Cambridge Chalk). By Prof. E. Forbes and J. W. Salter.

1866. Decade xii. (Plates 9 and 10). Fish from the Kimeridge Clay, Cottenham. By Prof. T. H. Huxley.

1872. The Geology of the London Basin. Part i. The Chalk and Eocene Beds of the Southern and Western Tracts (p. 45). By W. Whittaker.

1875. The Geology of Rutland and the Parts...of Cambridge included in Sheet 64. By J. W. Judd.

1877. The Geology of the Fenland. By S. B. J. Skertchly.

1878. The Geology of the N.W. part of Essex and the N.E. part of Herts with parts of Cambridgeshire and Suffolk (Explanation of Sheet 47). By W. H. Penning, W. H. Dalton, and F. J. Bennett.

1878. A Catalogue of the Cretaceous Fossils in the Museum of Practical Geology.

1878. Monograph iv. The Chimæroid Fishes of the British Cretaceous Rocks. By E. T. Newton.

1881. The Geology of the Neighbourhood of Cambridge (Explanation of Quarter Sheet 51 S.W. with part of 51 N.W.). By W. H. Penning and A. J. Jukes Browne.

1886. The Geology of the Country between and south of Bury St Edmunds and Newmarket (Explanation of Quarter Sheet 51 S.E.). By F. J. Bennett and J. H. Blake, edited by W. Whitaker.

1889. The Geology of London, Vol. I. (p. 46). By W. Whitaker.

1891. The Geology of Parts of Cambridgeshire and Suffolk (Ely, Mildenhall, and Thetford) (Explanation of Quarter Sheet 51 N.E. with part of 51 N.W.). By W. Whitaker (editor), H. B. Woodward, F. J. Bennett, S. B. J. Skertchly, and A. J. Jukes Browne.

1893. The Geology of South-Western Norfolk and of Northern Cambridgeshire (Explanation of Sheet 65). By W. Whitaker (editor), S. B. J. Skertchly, and A. J. Jukes Browne.

1897. The Jurassic Rocks of Britain. Vol. v. The Middle and Upper Oolitic Rocks of England. By H. B. Woodward.

BOOKS, PAPERS, ETC.

ADAMS, A. LEITH. **1879.** Monograph on the British Fossil Elephants (Cambridgeshire, p. 120). *Palæontograph. Soc.*

ANON. **1819.** Organic Remains (Antlers, Brink). *Quart. Journ. of Lit. Sci. and Arts*, vol. VII., p. 192.

    1836. A Notice of the Occurrence of certain Bodies in the Greensand at Cambridge, that are similar to those found in the Gault at Folkestone...and some Information on the Greensand and contiguous Strata at Cambridge. *Mag. Nat. Hist.*, vol. IX., p. 264.

    1862. Turtle Remains in the Upper Greensand. *Geologist*, vol. v., p. 73.

    1872. Excursion to Cambridge. *Proc. Geol. Assoc.*, vol. II., No. 5, p. 219.

BARRETT, L. **1859.** Geological Map of the Neighbourhood of Cambridge. (Drawn and coloured on the Ordnance Sheet 51 S.W.)

BARROIS, Dr C. **1876**. Recherches sur le Terrain Crétacé supérieur de l'Angleterre et de l'Irlande. *Mem. Soc. Geol. Nord,*
pp. 234. Abstract in *Ann. Soc. Geol. Nord,* t. III., p. 189.

BAYNE, A. D. **1872**. Royal Illustrated History of Eastern England,
including a Survey of the Eastern Counties, Physical Features,
Geology, etc. of Cambridgeshire, etc. Vol. I., 8vo., Yarmouth.

BELL, A. **1872**. *Unio limosus,* Nilsson, in the Crag [? Gravel].
Geol. Mag., vol. IX., p. 431.

BELL, Prof. **1863**. A Monograph of the Fossil Malacostracous
Crustacea of Great Britain. Part II. Crustacea of the Gault
and Greensand (Cambridge, pp. 4 etc.). *Palæontograph. Soc.*

BIDWELL, C. **1874**. Coprolites. *Trans. Instit. Surv.,* vol. VI., p. 293.

BLAKE, Rev. J. F. (and W. H. HUDLESTON). **1877**. On the
Corallian Rocks of England. *Quart. Journ. Geol. Soc.,* vol.
XXXIII., p. 260.

    1878. The Coral Rag of Upware. *Geol. Mag.,* dec. ii.,
vol. v., p. 90.

    1881. On the Correlation of the Upper Jurassic Rocks of
England with those of the Continent. Part I. *Quart. Journ.
Geol. Soc.,* vol. XXXVII., p. 497.

BONNEY, Prof. T. G. **1872**. Notes on the Roslyn Hill Clay Pit.
Geol. Mag., vol. IX., p. 403; Ditto, *Proc. Camb. Phil. Soc.,*
Part XIII., pp. 268, 269.

    1873. On the Upper Greensand or Chloritic Marl of
Cambridgeshire. *Proc. Geol. Assoc.,* vol. III., no. 1, p. 1.

    1875. Cambridgeshire Geology. 8vo., Cambridge.

    1877. Corallian of Upware. *Geol. Mag.,* dec. ii., vol. IV.,
p. 476.

BRADY, H. B. (and Dr W. B. CARPENTER). **1870**. Description
of *Parkeria* and *Loftusia,* two gigantic types of Arenaceous
Foraminifera. *Phil. Trans.,* vol. CLIX. (part 2), p. 721.

BRADY, G. S., Rev. H. W. CROSSKEY, and D. ROBERTSON. **1874**. A
Monograph of the Post-tertiary Entomostraca of Scotland
including Species from England and Ireland (Whittlesea, p. 108).
Palæontograph. Soc.

BRODIE, Rev. P. **1844**. Notice on the Occurrence of Land and
Freshwater Shells with Bones of some extinct Animals in the

Gravel near Cambridge. With notes by the Rev. Prof.
SEDGWICK. *Trans. Camb. Phil. Soc.*, vol. VIII., p. 1, p. 138.

1872. On Phosphatic and Bone-bed deposits in British
Strata, their economic uses and fossil contents. *36th Ann.
Rep. Warwick Nat. Hist. and Archæol. Soc.*, p. 53.

BRYLINSKI, M. (and G. LIONNET). 1878. Phosphates de Chaux
fossiles, Géologie et Origine, Applications au Agriculture.
Bull. Soc. Géol. Norm., t. IV., p. 3; *Phosphorites du Cambridge*,
pp. 92—97.

CARPENTER, Dr W. B. (and H. B. BRADY). 1870. Description of
Parkeria and *Loftusia*, two gigantic types of Arenaceous
Foraminifera. *Phil. Trans.*, vol. CLIX. (part 2), p. 721.

CARTER, Dr J. 1846. On the Occurrence of a new species of
Ichthyosaurus in the Chalk (Cambridge) (British Assoc.).
London Geol. Journ., p. 7.

1874. On a Skull of *Bos primigenius* perforated by a Stone
Celt. *Geol. Mag.*, dec. ii., vol. I., p. 492.

1886. On the Decapod Crustaceans of the Oxford Clay.
Quart. Journ. Geol. Soc., vol. XLII., p. 542.

1889. On Fossil Isopods, with a description of a new
Species. *Geol. Mag.*, dec. iii., vol. VI., p. 193.

CARTER, H. J. 1876. On the *Polytremata* (note on *Parkeria*).
Ann. Mag. Nat. Hist., ser. 4, vol. XVII., p. 208.

1877. On the close relationship of *Hydractinia*, *Parkeria*
and *Stromatopora*. *Ann. Mag. Nat. Hist.*, ser. 4, vol. XIX.,
p. 55.

1888. On two New Genera allied to *Loftusia* from the
Karakoram Pass and the Cambridge Greensand respectively.
Ann. Mag. Nat. Hist., ser. 6, vol. I., p. 172.

COLE, GRENVILLE A. J. 1895. Open-Air Studies. *London* (The
Fens, pp. 121—124).

CRADDOCK, T. (and N. WALKER). 1849. The History of Wisbech
and the Fens. Chap. I., "Physical Characteristics." Appendix, "Sketch of the Geology of the Fens," by H. M. LEE,
pp. 541—3. 8vo., Wisbech.

DAVIDSON, T. (and Prof. J. MORRIS). 1847. Description of some
species of Brachiopoda (Cambridge, p. 254). *Ann. Mag. Nat.
Hist.*, vol. XX., p. 250.

1852. A Monograph of British Cretaceous Brachiopoda (Cambridge, p. 42). *Palæontograph. Soc.*

1869. Notes on Continental Geology and Palæontology [Remarks on the Cambridge Greensand, with a letter from J. F. Walker]. *Geol. Mag.*, vol. VI., p. 259.

1874. A Monograph of the British Fossil Brachiopoda. Vol. IV., part 1. Supplement to the...Cretaceous Species (Cambridgeshire, p. 27, etc.). *Palæontograph. Soc.*

1884. Appendix to the Supplements of the Monograph on the British Fossil Brachiopoda. Vol. V. (Cambridgeshire, pp. 244, 247, etc.). *Palæontograph. Soc.*

DAVIES, W., and C. B. ROSE. **1864.** [Letters] on the Occurrence of Cycloid Fish-Scales, etc....in the Oolitic Formation. *Geol. Mag.*, vol. I., p. 92.

DAVIS, Prof. W. M. **1895.** The Development of certain English Rivers. *Geograph. Journ.*, vol. V., p. 127.

DENNIS, Rev. J. B. P. **1861.** On the Mode of Flight of the Pterodactyles of the Coprolite Bed near Cambridge. *Rep. Brit. Assoc.* for 1860; *Trans. of Sections*, p. 76.

DUNCAN, Prof. P. M. **1869.** First Report on the British Fossil Corals (Cambridge, p. 96). *Rep. Brit. Assoc.* for 1868, p. 75.

1869. A Monograph of the British Fossil Corals. Second Series (Cretaceous) (Cambridge, pp. 19, 20). *Palæontograph. Soc.*

FISHER, M. **1820.** Note on the Occurrence of the Bones of a Beaver, etc., near Ely. *Zoologist*, vol. I., p. 348.

FISHER, Rev. O. **1866.** On the Warp (of Mr Trimmer), its age, and probable connection with the last Geological Events. *Quart. Journ. Geol. Soc.*, vol. XXII., p. 553.

1867. On Roslyn or Roswell Hill Clay-pit, near Ely. *Proc. Camb. Phil. Soc.*, part IV., p. 51.

1868. Ditto. *Geol. Mag.*, vol. V., pp. 407, 438.

1871. On supposed Thermal Springs in Cambridgeshire [Letter]. *Geol. Mag.*, vol. VIII., p. 42.

1871. On Phenomena connected with Denudation, observed in the so-called Coprolite Pits near Haslingfield, Cambridgeshire. *Geol. Mag.*, vol. VIII., p. 65, and *Proc. Camb. Phil. Soc.*, part XII., pp. 195, 196.

1873. On the Phosphatic Nodules of the Cretaceous Rock of Cambridgeshire. *Quart. Journ. Geol. Soc.*, vol. XXIX., p. 52.

1879. On a Mammaliferous Deposit at Barrington near Cambridge. *Quart. Journ. Geol. Soc.*, vol. XXXV., p. 670.

FITTON, Dr W. H. **1836.** Observations on some of the Strata between the Chalk and the Oxford Oolite in the South-east of England. *Trans. Geol. Soc.*, ser. 2, vol. IV., p. 100. Abstract under different title, *Proc. Geol. Soc.*, vol. I., p. 26 (1827).

FORDHAM, H. G. **1874.** Notes on the Structure sometimes developed in Chalk. *Quart. Journ. Geol. Soc.*, vol. XXX., p. 43.

GARDNER, J. S. **1877.** On British Cretaceous *Patellidae* and other Families of Patelloid Gasteropoda. *Quart. Journ. Geol. Soc.*, vol. XXXIII., p. 192 (Cambridge, pp. 201, 202).

1878. On the Cretaceous *Dentaliidae*. *Quart. Journ. Geol. Soc.*, vol. XXXIV., p. 56.

1884. British Cretaceous *Nuculidae*. *Quart. Journ. Geol. Soc.*, vol. XL., p. 120.

1886. On Fossil Flowering or Phanerogamous Plants. *Geol. Mag.*, dec. iii., vol. 3, p. 495.

GEIKIE, Prof. JAMES. **1894.** The Great Ice Age. 3rd edition, pp. 342—352.

GRIFFITH, A. F. **1878.** On a Flint Implement from the Barnwell Gravel. *Geol. Mag.*, dec. ii., vol. v., p. 400.

HAILSTONE, Rev. Prof. J. **1815.** Supplementary Communication on Cambridge (Geol. Soc.). *Ann. of Phil.*, vol. v., p. 390.

1816. Outlines of the Geology of Cambridgeshire. *Trans. Geol. Soc.*, vol. iii., p. 243.

HAIME, J. and Prof. A. MILNE EDWARDS. **1850.** A Monograph of the British Fossil Corals. Part I. (Cambridge, pp. 63, etc.). *Palæontograph Soc.*

HAMILTON, W. J. **1850.** On the Occurrence of a Freshwater Bed of Marl in the Fens of Cambridgeshire. *Quart. Journ. Geol. Soc.*, vol. VI., p. 451.

HARKER, A. **1894.** Norwegian Rocks in the English Boulder Clays. *Geol. Mag.*, dec. iv., vol. I., p. 334.

HARMER, F. W. **1871.** On some Thermal Springs in the Fens of

Cambridgeshire. *Rep. Brit. Assoc.* for 1870, *Trans. of Sections*, p. 74.

    1871. The supposed Thermal Springs in Cambridgeshire [Letter]. *Geol. Mag.*, vol. VIII., p. 143.

HARRISON, W. J. **1882.** Geology of the Counties of England and of North and South Wales, pp. 20—28. London.

HENSLOW, Rev. Prof. **1846.** On Nodules, apparently Coprolitic, from the Red Crag, London Clay, and Greensand. *Rep. Brit. Assoc.* for 1845, *Trans. of Sections*, p. 51.

HILL, W. **1886.** On the Beds between the Upper and Lower Chalk of Dover and their comparison with the Middle Chalk of Cambridgeshire. *Quart. Journ. Geol. Soc.*, vol. XLII., p. 232.

HILL, W. (and A. J. JUKES-BROWNE). **1886.** The Melbourn Rock and the Zone of *Belemnitella plena* from Cambridge to the Chiltern Hills. *Quart. Journ. Geol. Soc.*, vol. XLII., p. 216.

    1887. On the Lower Part of the Upper Cretaceous Series in West Suffolk and Norfolk. *Quart. Journ. Geol. Soc.*, vol. XLIII., p. 544.

    1895. On the Occurrence of Radiolaria in Chalk. *Quart. Journ. Geol. Soc.*, vol. LI., p. 600.

HOWORTH, Sir Henry. **1887.** The Mammoth and the Flood. p. 110. 8vo. London.

    1892. The Mammoth and the Glacial Drift. *Geol. Mag.*, dec. iii., vol. IX., p. 399.

    1895. The Chalky Clay of the Fenland and its borders, its Constitution, Origin, Distribution and Age. *Q. J. G. S.*, vol. LI., p. 504.

    1896. The Destruction of the Chalk of Eastern England. *Geol. Mag.*, dec. iv., vol. III., p. 58.

    The Dislocation of the Chalk of Eastern England. *Ibid.*, p. 298.

    The Chalky and other Post Tertiary Clays of Eastern England. *Ibid.*, p. 449.

    The Middle Sands and Glacial Gravels of Eastern England. *Ibid.*, p. 533.

    1897. On the Erratic Boulders in the Drift of Eastern England. *Geol. Mag.*, dec. iv., vol. IV., pp. 123, 153.

    Water *versus* Ice. *Ibid.*, p. 213.

HUGHES, J. **1875**. Note on the Analysis of Cambridge Coprolite. *Chem. News*, vol. XXXI., p. 209.

HUGHES, Prof. T. MCKENNY. **1883**. On some Fossils supposed to have been found in the Pleistocene Gravels of Barnwell, near Cambridge. *Geol. Mag.*, dec. ii., vol. X., p. 454.

—— **1884**. Report of an excursion of the Geologists' Association to Cambridge, June 2 and 3, 1884. *Proc. Geol. Assoc.*, vol. VIII., no. 7, p. 399.

HUGHES, Mrs MCKENNY. **1888**. On the Mollusca of the Pleisto-cene Gravels in the neighbourhood of Cambridge. *Geol. Mag.*, dec. iii., vol. V., p. 193.

HUDLESTON, W. H. (and Rev. Prof. J. F. BLAKE). **1877**. On the Corallian Rocks of England. *Quart. Journ. Geol. Soc.*, vol. XXXIII., p. 260 (Cambridgeshire, pp. 313—315, 398).

—— **1878**. The Coral Rag of Upware. *Geol. Mag.*, dec. ii., vol. V., p. 90.

HUDLESTON, W. H. (and E. WILSON). **1892**. Catalogue of British Jurassic Gasteropoda. London.

JENYNS, Rev. L. **1846**. On the Turf of the Cambridgeshire Fens. *Rep. Brit. Assoc.* for 1845, *Trans. of Sections*, p. 75.

—— **1867**. A lecture on the Phosphatic Nodules obtained in the Eastern Counties, and used in Agriculture. *Proc. Bath Nat. Hist. and Antiq. Field Club*, vol. I., no. 1, p. 9.

JOHNSON, H. M. **1874**. On the Microscopic Structure of Flints and allied bodies. *Journ. Quek. Club*, vol. III., p. 234.

—— **1874**. The Nature and Formation of Flint and allied bodies. Pp. 16, 8vo., London.

JONAS, S. **1847**. On the Farming of Cambridgeshire (Remarks on the Geology, with Sections). *Journ. Roy. Agric. Soc.*, vol. VII., p. 35.

JONES, Prof. T. R. **1857**. A Monograph of the Tertiary Ento-mostraca of England (Cambridge, pp. 12, 14 etc.). *Palæonto-graph. Soc.*

JUKES-BROWNE, A. J. **1874**. Geological Map of the Neighbourhood of Cambridge. (Drawn and Coloured on the Ordnance Sheet 51 S.W.)

1875. On the Relations of the Cambridge Gault and Greensand. *Quart. Journ. Geol. Soc.*, vol. XXXI., p. 256.

1877. Supplementary Notes on the Fauna of the Cambridge Greensand. *Quart. Journ. Geol. Soc.*, vol. XXXIII., p. 485.

1878. The Post Tertiary Deposits of Cambridgeshire (Sedgwick Essay). 8vo., Cambridge and London.

1880. The Subdivisions of the Chalk. *Geol. Mag.*, dec. ii., vol. VII., p. 248.

1886. Handbook of Historical Geology. *London* (Cambridge, pp. 352, 390, 412).

1887. Note on a bed of Red Chalk in the Lower Chalk of Suffolk. *Geol. Mag.*, dec. iii., vol. IV., p. 24.

1893. The Amount of Disseminated Silica in Chalk considered in Relation to Flints. *Geol. Mag.*, dec. iii., vol. X., p. 541.

JUKES-BROWNE, A. J. (and W. HILL). **1886.** The Melbourn Rock and the Zone of *Belemnitella plena* from Cambridge to the Chiltern Hills. *Quart. Journ. Geol. Soc.*, vol. XLII., p. 216.

1887. On the Lower Part of the Upper Cretaceous Series in West Suffolk and Norfolk. *Quart. Journ. Geol. Soc.*, vol. XLIII., p. 544.

1895. On the Occurrence of Radiolaria in Chalk. *Quart. Journ. Geol. Soc.*, vol. LI., p. 600.

JUKES-BROWNE, A. J. (and W. J. SOLLAS). **1873.** On the Included Rock-fragments of the Cambridge Upper Greensand. *Quart. Journ. Geol. Soc.*, vol. XXIX., p. 11.

KEEPING, H. **1868.** Discovery of Gault with Phosphatic Stratum at Upware. *Geol. Mag.*, vol. V., p. 272.

KEEPING, Prof. W. **1883.** The Fossils and Palæontological Affinities of the Neocomian Deposits of Upware and Brickhill (Sedgwick Essay). 8vo., Cambridge.

LAKE, P. **1885.** On a Peculiar Form of *Hippopotamus major*, found at Barrington. *Geol. Mag.*, dec. iii., vol. II., p. 318.

LANKESTER, Prof. E. R. **1870.** On a new large *Terebratula* occurring in East Anglia. *Geol. Mag.*, vol. VII., p. 410.

LEIGHTON, T. (and J. E. MARR). **1894.** Excursion to Cambridge and Ely. *Proc. Geol. Assoc.*, vol. XIII., p. 292.

LEWIS, Prof. H. CARVILL. **1894.** The Glacial Geology of Great Britain and Ireland (edited by W. Crosskey), pp. 42 et seq., and pp. 328 et seq.

LIONNET, G. (and M. BRYLINSKI). **1878.** Phosphates de Chaux Fossiles, Géologie et Origine, Applications aux Agriculture. *Bull. Soc. Géol. Norm.*, t. IV., p. 3 ; *Phosphorites du Cambridge*, pp. 92—97.

LUNN, FRANCIS. **1818.** On the Strata of the Northern Division of Cambridgeshire. *Trans. Geol. Soc.*, vol. V., ser. 1, p. 114.

LYCETT, J. **1875.** Monograph of the British Fossil *Trigoniae.* No. III., p. 93. *Palæontograph. Soc.*

LYDEKKER, R. **1885-1887.** Catalogue of the Fossil Mammalia in the British Museum. Parts I—V. 8vo., London.

    1888-1890. Catalogue of the Fossil Reptilia and Amphibia in the British Museum. Parts I—IV. 8vo., London.

    1888. Note on the Classification of the Ichthyopterygia (with a notice of two new Species). *Geol. Mag.*, dec. iii., vol. V., p. 309.

    1889. Notes on the Remains and Affinities of five genera of Mesozoic Reptiles. *Quart. Journ. Geol. Soc.*, vol. XLV., p. 41.

    1889. On Remains of Eocene and Mesozoic Chelonia. *Quart. Journ. Geol. Soc.*, vol. XLV., p. 227.

    1891. Catalogue of the Fossil Birds in the British Museum. 8vo., London.

M'COY, Prof. F. **1848.** On some new Mesozoic Radiata. *Ann. Mag. Nat. Hist.*, ser. 2, vol. ii., p. 397.

    1849. On the Classification of some British Fossil Crustacea, etc. (Cambridge, p. 332). *Ann. Mag. Nat. Hist.*, ser. 2, vol. IV., p. 330.

    1854. On some new Cretaceous Crustacea (Cambridge, pp. 118, 120, 122). *Ann. Mag. Nat. Hist.*, ser. 2, vol. XIV., p. 116.

MARR, J. E. and T. LEIGHTON. **1894.** Excursion to Cambridge and Ely. *Proc. Geol. Assoc.*, vol. XIII., p. 292.

MARSHALL, W. **1874.** [Skulls from the Peat of the Isle of Ely.] *Journ. Anthrop. Instit.*, vol. III., p. 497.

MILLER, S. H., and S. B. J. SKERTCHLY. **1878.** The Fenland;
Past and Present. *Wisbech.*

MILNE-EDWARDS, Prof. A., and J. HAIME. **1850.** A Monograph of
the British Fossil Corals (Cambridge, pp. 63, etc.). *Palæontograph. Soc.*
        1868. Note on the existence of a large Pelican in the
Turbaries of England (Transl. from *Comptes Rendus*, p. 1242).
Ann. Mag. Nat. Hist., ser. 4, vol. ii., p. 165.

MITCHELL, Dr J. **1838.** On the Drift from the Chalk and the
Strata below the Chalk in the Counties of Norfolk, Suffolk,
Essex, Cambridge, etc. *Proc. Geol. Soc.*, vol. III., p. 3.

MOORE, N. **1867.** *Megaceros hibernicus* in the Cambridgeshire
Fens. *Ann. Mag. Nat. Hist.*, ser. 3, vol. XX., pp. 77, 301.

MORRIS, Prof. J., and T. DAVIDSON. **1847.** Description of some
species of Brachiopoda (Cambridge, p. 254). *Ann. Mag. Nat.
Hist.*, vol. XX., p. 250.

MORRIS, Prof. JOHN. **1883.** The Chalk; its Distribution and
Subdivisions. *Proc. Geol. Assoc.*, vol. VIII., p. 208.

NEWTON, E. T. **1888.** Notes on Pterodactyls. *Proc. Geol. Assoc.*,
vol. X., p. 406.

NICHOLSON, Prof. H. A. **1888.** On the Structure and Affinities of
the genus *Parkeria. Ann. Mag. Nat. Hist.*, ser. 6, vol. I., p. 1.

OKES, J. **1821.** An Account of some Fossil Remains of the
Beaver, found in Cambridgeshire. *Trans. Camb. Phil. Soc.*,
vol. I., p. 175.

OWEN, Prof. R. **1840.** Report on British Fossil Reptiles (Cambridgeshire, pp. 74, 75). *Rep. Brit. Assoc.* for 1839, p. 43.
        1842. Report on British Fossil Reptiles (Part 2, Cambridge,
p. 172). *Rep. Brit. Assoc.* for 1841, p. 60.
        1843. Report on the British Fossil Mammalia (Cambridge,
pp. 64, 69). *Rep. Brit. Assoc.* for 1842, p. 54.
        1846. A History of British Fossil Mammals and Birds.
8vo., London (Cambridge, pp. 105, 119, 195, etc.).
        1851. Monograph on the Fossil Reptilia of the Cretaceous
Formations (Cambridge, pp. 8, 19, 55, 64, 72, etc.). *Palæontograph. Soc.*

1859. On Remains of New and Gigantic Species of Pterodactyle (*P. Fittoni* and *P. Sedgwicki*) from the Upper Greensand near Cambridge. *Rep. Brit. Assoc.* for 1858 ; *Trans. of Sections*, p. 98.

1859. On the Vertebral Characters of the Order Pterosauria, as exemplified in the Genera *Pterodactylus* and *Dimorphodon*. *Phil. Trans.*, vol. CXLIX. (part 1), p. 161.

1859. Monograph on the Fossil Reptilia of the Cretaceous Formation. Supplement No. 1, Pterosauria (*Pterodactylus*). *Palæontograph. Soc.*

1861. Monograph of the Fossil Reptilia of the Cretaceous and Purbeck Strata (Supplement). *Palæontograph. Soc.*

1864. Monograph of the Fossil Reptilia of the Cretaceous and Purbeck Strata (Supplement). *Palæontograph. Soc.*

PENNING, W. H. **1876.** Notes on the Physical Geology of East Anglia during the Glacial Period. *Quart. Journ. Geol. Soc.*, vol. XXXII., p. 191.

PHILLIPS, Prof. J. **1870.** A Monograph of British Belemnitidae. Part v.

PORTER, Dr H. **1861.** The Geology of Peterborough and its neighbourhood. 8vo., Peterborough.

1863. On the Occurrence of large Quantities of Fossil Wood in the Oxford Clay near Peterborough. *Quart. Journ. Geol. Soc.*, vol. XIX., p. 317.

PRESTWICH, Prof. Sir J. **1851.** A Geological Inquiry respecting the Water-bearing Strata of the Country around London, etc. 8vo., London (Cambridge, pp. 76, 90, 144, 146, 167, 183, 192, 193, 235).

PRICE, F. G. H. **1879.** The Gault, being the Substance of a Lecture delivered in the Woodwardian Museum, Cambridge. 8vo., London.

REID, W. C. **1876.** Mineral Phosphates and Superphosphate of Lime. *Chem. News*, vol. XXXIV., pp. 48—50.

ROBERTS, T. **1887.** On the Correlation of the Upper Jurassic Rocks of the Swiss Jura with those of England. *Quart. Journ. Geol. Soc.*, vol. XLIII., p. 229.

1891. On two abnormal Cretaceous Echinoids. *Geol. Mag.*, dec. iii., vol. VIII., p. 116.

1892. The Jurassic Rocks of the Neighbourhood of Cambridge. Sedgwick Essay, 8vo., *Cambridge.*

ROSE, C. B. **1859.** Geological Pearls (Ely, Cherryhinton). *Geologist*, vol. II., p. 295.

ROSE, C. B. (and W. DAVIES). **1864.** [Letters] on the Occurrence of Cycloid Fish-scales, etc.,...in the Oolitic Formation. *Geol. Mag.*, vol. I., p. 92.

ROWE, Rev. A. W. **1887.** Rocks of the Essex Drift. *Quart. Journ. Geol. Soc.*, vol. XLIII., p. 351.

SEDGWICK, Prof. A. **1825.** On the Origin of Alluvial and Diluvial Formations (Cambridge, pp. 244, 251 and 22). *Ann. of Phil.*, ser. 2, vol. IX., p. 241, and vol. X., p. 18.

1844. Notes to a Paper by Rev. P. B. Brodie on the Occurrence of Land and Freshwater Shells with Bones of some extinct Animals in the Gravel near Cambridge. *Trans. Camb. Phil. Soc.*, vol. VIII., pt. 1, p. 138.

1846. On the Geology of the Neighbourhood of Cambridge, including the Formations between the Chalk Escarpment and the Great Bedford Level. *Rep. Brit. Assoc.* for 1845, *Trans. of Sections*, p. 40.

1861. A Lecture on the Strata near Cambridge and the Fens of the Bedford Level. 8vo. (Privately printed.)

1869. Prefatory Note to Prof. H. G. Seeley's Index to the Fossil Remains of Aves, etc. in the Woodwardian Museum, Cambridge.

SEELEY, Prof. H. G. **1861.** The Fen Clay Formation (Camb. Phil. Soc., Oct. 28). *The Geologist*, p. 552 ; *Ann. Mag. Nat. Hist.*, ser. 3, vol. VIII., p. 365.

1861. Notes on Cambridge Palæontology. No. 1, Some new Upper Greensand Bivalves ; No. 2, Some new Gasteropods from the Upper Greensand ; No. 3, On a new order of Echinoderms ; No. 4, Some new Upper Greensand Echinoderms. *Ann. Mag. Nat. Hist.*, ser. 3, vol. VII., pp. 116, 281, 365, and vol. VIII., p. 16.

1861. On some Anomalous Fossils from the Upper Greensand of Cambridge. *Proc. Geol. Assoc.*, vol. I., p. 147.

1862. Notes on Cambridge Geology. 1, Preliminary Notice of the Elsworth Rock and associated Strata (Brit. Assoc.). *Ann. Mag. Nat. Hist.*, ser. 3, vol. x., p. 97.

1863. On a Whittled Bone from the Barnwell Gravel. *Rep. Brit. Assoc.* for 1862, *Trans. of Sections*, p. 94.

1864. A Monograph of the Ammonites of the Cambridge Greensand. *Quart. Journ. Geol. Soc.*, vol. xx., p. 166.

1864. *Mytilus spathulatus*, a new Cretaceous Species. *Geologist*, vol. vii., p. 53.

1864. On a section of the Lower Chalk near Ely. *Geol. Mag.*, vol. i., p. 150.

1865. On a new Lizard with Ophidian Affinities from the Lower Chalk (*Saurospondylus dissimilis*). *Ann. Mag. Nat. Hist.*, ser. 3, vol. xvi., p. 105.

1865. On Ammonites from the Cambridge Greensand. *Ann. Mag. Nat. Hist.*, ser. 3, vol. xvi., p. 225.

1865. On the Fossil Neck-bones of a Whale from the neighbourhood of Ely. *Geol. Mag.*, vol. ii., p. 54.

1865. On the significance of the sequence of Rocks and Fossils. Theoretical considerations of the Upper Secondary Rocks, as seen in the section at Ely. *Geol. Mag.*, vol. ii., p. 262.

1865. On a Section discovering the Cretaceous Rocks at Ely. *Geol. Mag.*, vol. ii., p. 529.

1866. A Sketch of the Gravels and Drift of Fenland. *Quart. Journ. Geol. Soc.*, vol. xxii., p. 470.

1866. The Rock of the Cambridge Greensand. *Geol. Mag.*, vol. iii., p. 302.

1866. Theoretical Remarks on the Gravel and Drift of the Fenlands. *Geol. Mag.*, vol. iii., p. 495.

1867. On the Association of Potton Sand Fossils with those of the Farringdon Gravels in a phosphatic deposit at Upware on the Cam. *Proc. Camb. Phil. Soc.*, Parts v., vi., p. 99.

1868. On the Collocation of the Strata at Roswell Hole near Ely. *Geol. Mag.*, vol. v., p. 347.

1869. Discovery of *Dakosaurus* in England. *Geol. Mag.*, vol. vi., p. 188.

1869. Index to the Fossil Remains of Aves, Ornithosauria, and Reptilia from the Secondary System of Strata, arranged in

the Woodwardian Museum of the University of Cambridge.
With a Prefatory Note by the Rev. Prof. Sedgwick. 8vo.,
Cambridge and London.

1870. The Ornithosauria: an Elementary Study of the
Bones of Pterodactyles, made from Fossil Remains found in the
Cambridge Upper Greensand. 8vo., Cambridge and London.

1870. On the Frontal Bone in the Ornithosauria, with
additional evidence of the Structure of the Head in Pterodactyles from the Cambridge Upper Greensand. *Proc. Camb.
Phil. Soc.*, No. XI., p. 186 (8 lines).

1870. Remarks on Prof. Owen's Monograph on *Dimorphodon. Ann. Mag. Nat. Hist.*, ser. 4, vol. VI., p. 129.

1871. Additional Evidence of the Structure of the Head in
Ornithosaurs from the Cambridge Upper Greensand : being a
Supplement to the "Ornithosauria." *Ann. Mag. Nat. Hist.*,
ser. 4, vol. VII., p. 20.

1871. On *Acanthopholis platypus* (Seeley), a Pachypod
from the Cambridge Upper Greensand. *Ann. Mag. Nat. Hist.*,
ser. 4, vol. VIII., p. 305.

1873. On *Cetarthrosaurus Walkeri* (Seeley), an Ichthyo-
saurian from the Cambridge Upper Greensand. *Quart. Journ.
Geol. Soc.*, vol. XXIX., p. 505.

1874. On Cervical and Dorsal Vertebrae of *Crocodilus
cantabrigiensis* (Seeley), from the Cambridge Upper Greensand.
Quart. Journ. Geol. Soc., vol. XXX., p. 693.

1876. On an Associated Series of Cervical and Dorsal
Vertebrae of *Polyptychodon*, from the Cambridge Upper Green-
sand in the Woodwardian Museum of the University of
Cambridge. *Quart. Journ. Geol. Soc.*, vol. XXXII., p. 433.

1876. On *Crocodilus icenicus* (Seeley), a second and larger
species of Crocodile from the Cambridge Upper Greensand
contained in the Woodwardian Museum. *Quart. Journ. Geol.
Soc.*, vol. XXXII., p. 437.

1876. On *Macrurosaurus semnus* (Seeley), a long-tailed
animal with procœlous vertebrae from the Cambridge Upper
Greensand, preserved in the Woodwardian Museum. *Quart.
Journ. Geol. Soc.*, vol. XXXII., p. 440.

1876. On British Fossil Cretaceous Birds. *Quart. Journ.
Geol. Soc.*, vol. XXXII., p. 490.

1879. On the Dinosauria of the Cambridge Greensand. *Quart. Journ. Geol. Soc.*, vol. XXXV., p. 591.

1881. On the Evidence of Two Ornithosaurians referable to the Genus *Ornithocheirus*, from the Upper Greensand of Cambridge. *Geol. Mag.*, dec. ii., vol. VIII., p. 13.

1887. On *Patricosaurus merocratus*, a Lizard from the Cambridge Greensand. *Quart. Journ. Geol. Soc.*, vol. XLIII., p. 216.

1891. The Ornithosaurian Pelvis. *Ann. Mag. Nat. Hist.*, ser. 6, vol. VII., p. 237.

1891. On the Shoulder-Girdle in Cretaceous Ornithosauria. *Ann. Mag. Nat. Hist.*, ser. 6, vol. VII., p. 438.

SHARPE, D. **1855.** Description of the Fossil Remains of Mollusca found in the Chalk of England. Part 2, Cephalopoda (Cambridge, p. 29). *Palæontograph. Soc.*

SHERBORN, C. D., and A. SMITH WOODWARD. **1890.** A Catalogue of British Fossil Vertebrata. London.

SKERTCHLY, S. B. J., and S. H. MILLER. **1878.** The Fenland; Past and Present. Wisbech.

SMITH, W. **1819.** Geological View and Section of the Country between London and Cambridge.

1819. Geological View and Section through Suffolk to Ely.

SOLLAS, Prof. W. J. **1872.** Some Observations on the Upper Greensand Formation of Cambridge. *Quart. Journ. Geol. Soc.*, vol. XXVIII., p. 397.

1872. New British Crustacean (Upware). *Geol. Mag.*, vol. IX., p. 144.

1873. On the *Ventriculitae* of the Cambridge Upper Greensand. *Quart. Journ. Geol. Soc.*, vol. XXIX., p. 63.

1873. On the Coprolites of the Upper Greensand Formation and on Flints. *Quart. Journ. Geol. Soc.*, vol. XXIX., p. 76.

1873. On the Foraminifera and Sponges of the Upper Greensand of Cambridge. *Geol. Mag.*, vol. X., p. 268, and *Proc. Camb. Phil. Soc.*, Part XIV., pp. 299, 300.

1877. On *Pharetrospongia Strahani* (Sollas), a fossil Holoraphidote Sponge from the Cambridge "Coprolite" Bed. *Quart. Journ. Geol. Soc.*, vol. XXXIII., p. 242.

SOLLAS, Prof. W. J. (and A. J. JUKES-BROWNE). **1873.** On the

Included Rock-fragments of the Cambridge Upper Greensand. *Quart. Journ. Geol. Soc.*, vol. XXIX., p. 11.

SOWERBY, J. **1815.** The Mineral Conchology of Great Britain. Vol. I., p. 201. 8vo., London.

  1824–1825. The Mineral Conchology of Great Britain. Vol. V. (Cambridgeshire, pp. 5, 6, 53, 54). 8vo., London.

SPENCE, —. **1857.** [On Coprolites.] *Proc. Lit. and Phil. Soc. Manchester*, vol. I., no. 1, p. 3.

TALBOT, H. T. **1875.** The Chloritic Marl of Cambridgeshire. 3rd *Rep. Winchester Coll. Nat. Hist. Soc.*, p. 36.

TEALL, J. J. H. **1875.** The Potton and Wicken Phosphatic Deposits. Sedgwick Essay, 8vo., Cambridge.

THACKERAY, F. **1822.** On a Remarkable Instance of Fossil Organic Remains found near Streatham in the Isle of Ely· *Trans. Camb. Phil. Soc.*, vol. I., part ii., p. 459.

TOMES, R. F. **1883.** On some new or imperfectly known Madreporaria, from the Coral Rag of the Counties of Wilts, Oxford, Cambridge, and York. *Quart. Journ. Geol. Soc.*, vol. XXXIX., p. 555.

  1885. Observations on some imperfectly known *Madreporaria* from the Cretaceous Formation of England. *Geol. Mag.*, dec. iii., vol. II., p. 541.

TRIMMER, J. **1854.** On some Mammaliferous Deposits in the Valley of the Nene, near Peterborough. *Quart. Journ. Geol. Soc.*, vol. X., p. 343.

VANCOUVER, C. **1794.** General View of the Agriculture in the County of Cambridge. (Map of Soils. Appendix on the Fens, with Borings.) 4to., London.

VINE, G. R. **1885.** Notes on the Polyzoa and Foraminifera of the Cambridge Greensand. *Proc. Geol. Soc.*, p. 101.

VOELCKER, Dr A. **1860.** On the Chemical Composition and Commercial Value of...Coprolites and other Phosphatic Materials used in England for Agricultural Purposes (Analyses of Cambridge "Coprolites," pp. 357, 358). *Journ. Roy. Agric. Soc.*, vol. XXI., p. 350.

  1875. On the Chemical Composition of Phosphatic Minerals used for Agricultural Purposes. *Journ. Roy. Agric. Soc.*, ser. 2, vol. XI., p. 399.

WALKER, J. F. **1867.** On some new Coprolite Workings in the Fens. *Geol. Mag.*, vol. IV., p. 309.

1867. On some new *Terebratulidae* from Upware. *Geol. Mag.*, vol. IV., p. 454.

1868. On a new Phosphatic Deposit near Upware, Cambridgeshire. *Rep. Brit. Assoc.* for 1867 ; *Trans. of Sections*, p. 73.

1868. On the Species of Brachiopoda which occur in the Lower Greensand at Upware. *Geol. Mag.*, vol. V., p. 399.

1868. Occurrence of *Terebratula (Waldheimia) pseudojurensis* (Leymerie) in England [at Upware]. *Ann. Mag. Nat. Hist.*, ser. 4, vol. I., p. 386.

1868. On the Occurrence of the Genus *Anser* in the Peat and Gravel Deposits in Cambridgeshire. *Ann. Mag. Nat. Hist.*, vol. II., p. 388.

1870. On Secondary Species of Brachiopoda. *Proc. Camb. Phil. Soc.*, no. XI., p. 560.

WALKER, N. and T. CRADDOCK. **1849.** The History of Wisbech and the Fens. Chap. I., "Physical Characteristics"; Appendix, "Sketch of the Geology of the Fens," by H. M. Lee, pp. 541–3. 8vo., Wisbech.

WARBURTON, H. **1814.** A Description of some Specimens from the Neighbourhood of Cambridge. (*Geol. Soc.*) *Ann. of Phil.*, vol. III., p. 72.

WATSON, W. **1827.** Historical Account of Wisbeach (Notices of Bones, pp. 58, 578, etc.).

WATTS, W. W. **1880.** Pebble from the Cambridge Greensand. *Geol. Mag.*, dec. ii., vol. VIII., p. 95.

WELLS, W. **1860.** The Drainage of Whittlesea Mere (Analyses of Clays, etc., p. 148). *Journ. Roy. Agric. Soc.*, vol. XXI., p. 134.

WESTLAKE, E. **1888.** Tabular Index to the Upper Cretaceous Fossils of England and Ireland cited by Dr Ch. Barrois in his *Descr. Géol. de la Craie de l'Ile de Wight* (1875) and *Rech. sur le Terr. Crét. Sup. de l'Angleterre et de l'Irlande* (1876). Fordingbridge.

WHITAKER, W. **1890.** On a Deep Channel of Drift in the Valley of the Cam, Essex. *Quart. Journ. Geol. Soc.*, vol. XLVI., p. 333.

WILSON, E. and W. H. HUDLESTON. 1892. Catalogue of British Jurassic Gasteropoda. London.

WOOD, S. V., jun. 1865. A Map of the Upper Tertiaries in the Counties of Norfolk, Suffolk, Essex, Middlesex, Hertford, Cambridge, etc. (with Remarks and Sections). Privately printed.

    1867. On the Structure of the Post-glacial Deposits in the South-east of England, *Quart. Journ. Geol. Soc.*, vol. XXIII., p. 394 (corrections in *Geol. Mag.*, vol. v., pp. 43, 534).

    1876. Physical Geology of East Anglia in the Glacial Epoch. *Geol. Mag.*, dec. ii., vol. III., p. 284.

    1880. On the Newer Pliocene Period in England. Part 1. *Quart. Journ. Geol. Soc.*, vol. XXXVI., p. 457.

    1882. On the Newer Pliocene Period in England. Part 2. *Quart. Journ. Geol. Soc.*, vol. XXXVIII., p. 667.

WOODS, H. 1891. Catalogue of the Type-Fossils in the Woodwardian Museum, Cambridge, with a preface by Prof. T. McKenny Hughes. Cambridge.

    1893. Additions to the Type-Fossils in the Woodwardian Museum. *Geol. Mag.*, dec. iii., vol. x., p. 111.

    1896. The Mollusca of the Chalk Rock, Part 1. *Quart. Journ. Geol. Soc.*, vol. LII., p. 68.

    1897. *Ibid.*, Part 2. *Quart. Journ. Geol. Soc.*, vol. LIII., p. 377.

WOODWARD, B. B. 1888. Notes on the Pleistocene Land and Freshwater Mollusca from the Barnwell Gravels. *Proc. Geol. Assoc.*, vol. x., p. 355:

WOODWARD, Dr H. 1870. Contributions to British Fossil Crustacea. *Proc. Camb. Phil. Soc.*, p. 493.

    1885. On "Wingless Birds," Fossil and Recent. *Geol. Mag.*, dec. iii., vol. II., p. 308 ; Ditto, *Proc. Geol. Assoc.*, vol. IX., p. 352.

WOODWARD, H. B. 1887. The Geology of England and Wales. 2nd ed., 1887. *London* (Cambridgeshire, pp. 324, 329, 335, etc.).

WOODWARD, A. SMITH. 1885. On the Literature and Nomenclature of British Fossil Crocodilia. *Geol. Mag.*, dec. iii., vol. II., p. 496.

1887. A Synopsis of the Vertebrate Fossils of the English Chalk. *Proc. Geol. Assoc.*, vol. x., p. 273.

1889. On the so-called Cretaceous Lizard *Rhaphiosaurus*. *Ann. Mag. Nat. Hist.*, ser. 6, vol. iv., p. 350.

1889. Preliminary Notes on some new and little known British Jurassic Fishes. Brit. Assoc., 1889. *Geol. Mag.*, dec. iii., vol. vi., p. 448.

1890. On some British Jurassic Fish-remains referable to the Genera *Eurycormus* and *Hypsocormus*. *Proc. Geol. Soc.*, No. 8.

1890. On a head of *Eurycormus* from the Kimmeridge Clay of Ely. *Geol. Mag.*, dec. iii., vol. vii., p. 289.

WOODWARD, A. SMITH. **1893.** Some Cretaceous Pycnodont Fishes. *Geol. Mag.*, dec. iii., vol. x., pp. 433, 487.

1893. Notes on the Sharks' Teeth from British Cretaceous Formations. *Proc. Geol. Assoc.*, vol. xiii., p. 190.

1895. A Synopsis of the Remains of Ganoid Fishes from the Cambridge Greensand. *Geol. Mag.*, dec. iv., vol. ii., p. 207.

WOODWARD, A. S. (and C. D. SHERBORN). **1890.** A Catalogue of British Fossil Vertebrata. London.

WOODWARD, Dr S. P. **1864.** Note on *Plicatula sigillina*, an undescribed Fossil of the Upper Chalk and Cambridge Phosphate Bed. *Geol. Mag.*, vol. i., p. 112.

WRIGHT, Dr T. **1859.** A Monograph on the British Fossil Echinodermata from the Oolitic Formations. Part 3 (Cambridge, p. 318). *Palæontograph. Soc.*

1864. A Monograph of the Fossil Echinodermata from the Cretaceous Formations (Cambridge, p. 39). *Palæontograph. Soc.*

1874. Monograph on the British Fossil Echinodermata from the Cretaceous Formations. Vol. i., part 6, on the *Echinoconidae* (Cambridgeshire, pp. 208—210). *Palæontograph. Soc.*

WRIGHT, Dr G. F. **1892.** Man and the Glacial Period. (Internat. Scientif. Ser.), p. 158 et seq.

INDEX.

CAMBRIDGE: PRINTED BY J. AND C. F. CLAY, AT THE UNIVERSITY PRESS.

Printed in the United States
By Bookmasters